T0137207

Advanced Sciences and Technologies for Security Applications

The series Advanced Sciences and Technologies for Security Applications comprises interdisciplinary research covering the theory, foundations and domain-specific topics pertaining to security. Publications within the series are peer-reviewed monographs and edited works in the areas of:

- biological and chemical threat recognition and detection (e.g., biosensors, aerosols, forensics)
- crisis and disaster management
- terrorism
- cyber security and secure information systems (e.g., encryption, optical and photonic systems)
- traditional and non-traditional security
- energy, food and resource security
- economic security and securitization (including associated infrastructures)
- transnational crime
- human security and health security
- social, political and psychological aspects of security
- recognition and identification (e.g., optical imaging, biometrics, authentication and verification)
- smart surveillance systems
- applications of theoretical frameworks and methodologies (e.g., grounded theory, complexity, network sciences, modelling and simulation)

Together, the high-quality contributions to this series provide a cross-disciplinary overview of forefront research endeavours aiming to make the world a safer place.

The editors encourage prospective authors to correspond with them in advance of submitting a manuscript. Submission of manuscripts should be made to the Editor-in-Chief or one of the Editors.

Mohiuddin Ahmed · Paul Haskell-Dowland
Editors

Cybersecurity for Smart Cities

Practices and Challenges

 Springer

Editors
Mohiuddin Ahmed
School of Science
Edith Cowan University
Joondalup, WA, Australia

Paul Haskell-Dowland
School of Science
Edith Cowan University
Joondalup, WA, Australia

ISSN 1613-5113   ISSN 2363-9466 (electronic)
Advanced Sciences and Technologies for Security Applications
ISBN 978-3-031-24948-8   ISBN 978-3-031-24946-4 (eBook)
https://doi.org/10.1007/978-3-031-24946-4

This Springer imprint is published by the registered company Springer Nature Switzerland AG
The registered company address is: Gewerbestrasse 11, 6330 Cham, Switzerland

Dedicated to

My Loving Wife: Raiyan

Mohiuddin Ahmed

Preface

Ensuring cybersecurity for smart cities should not be an afterthought and is a considerable challenge. In addition, the security configuration of computing devices and software updates is often overlooked when it is most needed to fight cybercrime and ensure data privacy. Therefore, the threat landscape in the smart city context becomes wider and far more challenging. There is a clear need for collaborative work throughout the entire value chain of the network. In this context, this book will address the cybersecurity challenges associated with the smart cities, providing a bigger picture of the concepts, intelligent techniques, practices, and open research directions in this area.

Chapters

The book reflects several aspects of cyber smart cities, including the practices and challenges ahead. Unlike other books on similar topics, the book focuses on interesting aspects of cyber smart cities, such as bullying via social media, misinformation detection, data and utilities, ethical aspects, human resource, deep learning, healthcare ecosystem, smart vehicles, deep learning, and last but not least cryptocurrency followed by a national security threat, i.e., ransomware attacks. Chapter "Investigation on the Infusion of Cybersecurity and Smart City" explores how cybersecurity and smart cities cross paths. Chapter "Cyberbullying Instilled in Social Media" focuses on the weaponization of social media for cyberbullying. Chapter "Pivoting Human Resource Policy Around Emerging Invasive and Non-invasive Neurotechnology" dives into the psychological aspects of human resources regarding emerging technologies. Chapter "Privacy and Ethics in a Smart City: Towards Attaining Digital Sovereignty" touches on the privacy and ethical aspects in the realm of smart cities. Chapter "Knowledge Organization Systems to Support Cyber-Resilience in Medical Smart Home Environments" highlights cyber safety in smart home environments for better medical support. Chapters "Cybersafe Capabilities and Utilities for Smart Cities" and "Cyber Safe Data Repositories" highlight the challenges associated with data repositories and utilities of cyber smart cities. Chapter "Misinformation Detection in Cyber Smart Cities" showcases a very interesting aspect of today's Internet ecosystem, which is misinformation detection. Chapter "Vehicle Trajectory Obfuscation and Detection" highlights cyber issues in smart vehicles used in smart cities. Chapter "The Contribution of Deep Learning for Future Smart Cities" showcases the contribution of deep learning to smart cities. Chapter "An EDGE Supported Ambulance Management System for Smart Cities" dives into investigating the effectiveness of edge computing in supporting ambulances for smart cities. Chapters "Cryptocurrency: Is it the Future of Payments?" and "Ransomware: A Threat to Cyber Smart Cities" bring together the most pressing issues, i.e., cryptocurrency and ransomware attacks.

Joondalup, Australia   Mohiuddin Ahmed
                                                                          Paul Haskell-Dowland

Acknowledgements

It is an exciting book editing experience and our sincere gratitude to the publisher for facilitating the process. This book editing journey enhanced our patience, communication, and determination to contribute to the science community. We thank all the contributors, reviewers, and the publishing team. Lastly, we thank our family members whose support and encouragement contributed significantly to completing this book.

Mohiuddin Ahmed
Paul Haskell-Dowland

Contents

Investigation on the Infusion of Cybersecurity and Smart City

Shahrin Zubair, Moinuddin Zubair, and Mohiuddin Ahmed

Abstract In order to escalate the operational efficiency and improve the two-way quality-of-service between government and citizens, as well as sharing significant information with the general public by means of information and communication technologies (ICT) is what is referred to as a Smart City. Not every city in the world is yet a smart one but it is almost on the verge of becoming one by the absolute application of smart components. Smart technological components involve the personal and private data of any user which makes them risky and vulnerable as there is a higher risk of increase in several issues and challenges. One cyber-attack can vividly destroy all the security aspects of the smart city and put the whole arena at risk. There are several other cyber-threats that in completion can majorly affect the behavioral strategies of a smart city. Thus, to prevent the malicious manipulation of the use of data and infrastructure of smart cities, it is important to introduce cyber security in the same direction. The objective of this book chapter is to introduce the steps presently taken or are required to be taken to ensure cyber security in this particular technological area. We aim to focus on the following major smart components of a smart city: grid, transportation system, building, and healthcare and their relevant cyber-security aspects that need to be framed so as to assure safe prevention from serious security breach incidents.

Keywords Cyber smart city · Cyber security

S. Zubair (✉)
Student Connect BD, Chattogram, Bangladesh
e-mail: shahrinsadik.ss@gmail.com

M. Zubair
Department of Computer Science and Engineering, Brac University, Dhaka, Bangladesh

M. Ahmed
School of Science, Edith Cowan University, Perth, Australia

M. Ahmed and P. Haskell-Dowland (eds.), *Cybersecurity for Smart Cities*,
Advanced Sciences and Technologies for Security Applications,
https://doi.org/10.1007/978-3-031-24946-4_1

1 Introduction

Internet of Things (IoT) is defined as a network of connected physical objects with sensors and ability of processing data with several other technological devices in order to be able to connect and exchange data within each other [1]. Any particular thing or electrical device that has the capability to get connected to the internet and exchange data as a part of information comes under the umbrella of IoT. Several traditionally common but new disruption of IoT includes, Smart home, Smart healthcare system, automotive waste management system, mobile pharmacy, smart fire alarms, smart traffic management, smart security system, and smart city etc. [2]. Amongst all these, smart city is the updated version of a casual urban life which enables all the connected devices to share and transfer data within each other easing the as usual pattern of an individual's life in a city. In order to meet the social needs of a city, it is important and also required to implement a technologically advanced solution for the city by introducing Information and Communication Technologies (ICT) with an amalgamation of IoT [3]. IoT tends to contribute as a strategy to take an attempt to bring immense changes in a basic lifestyle of the residents. A smart city is absolutely dependent on the big data collected from the smart components around the city. It is made up of several IoT sensors, engines and people with the motive to provide an efficient operation of the daily tasks. ICTs are directly utilized as a means for enhancing performance, consistency and interactions of the connected components of the city which eventually improves the communication between the connected nodes and reduces the overall cost by consumption of resources where and when needed. The prime highlight of the philosophy of a smart city is to escalate the standards as well as establishing efficient and more reliable technological solutions within the diverse network services of the metropolitan arena. IoT is assumed to be the chief model in order to make an intelligent city as its ability to fabricate smart appliances within the city enables the communities to be quick on the uptake. One of the prominent and vital components of an IoT application such as a smart city, is the sensor which interconnects the related objects across the network so as to share the information and deliver vital services within the residents of a smart city. Though IoT is a combination of sensors, electronic devices and other relevant softwares to control and monitor the overall performance and efficacy of the other related parts of a system, there is still a higher risk of facing a challenge for any IoT driven company to constantly being able to generate these data. IoT [1] plays a key role in our lives and to adopt to it as well as using it has unexpectedly become a common mantra at present which ultimately enforces these connected items to be the significant components for future industry of Internet. IoT generates data, ideas and experiences with every passing day that eventually enables numerous devices to smooth the way for a better quality of our everyday life. With each passing day it has become a significant requirement to maintain a city's continual adaptability to respective social, ecological, technical and economical aspects with regular evolution of the residents. Due to a gradual growth in all aspects of a city [4], more and more concerning issues are being raised such as traffic congestion, environment pollution, and extreme demand for services like energy and sanitation. Thus, introducing smart components in a city predominantly is expected to resolve these issues as a part of a smart solution evolved from digitized services, automated processes and decisions made on the basis of gathered big data making it a smart city. Adopting such services enables a city to enhance the administrative and operative actions and challenges with the aim to produce a sustainable ambience for the residents. A smart city gives an opportunity to identify and decode the streamlined patterns of its community, economic and ecological aspects leading to an appropriate real-time decision making by the city stakeholders so as to maintain and control its sustainability and durability. As IoT based solutions are commonly increasing with time, there is a greater possibility of facing severe cyber-attacks on them proportionately rising to severe security and privacy issues.

The chapter is organized as follows: Sect. 3 Smart City, Sect. 4 The prominent characteristics of a Smart City, Sect. 5 Applications of a Smart City, Sect. 6 Cybersecurity and Smart city, Sect. 7 Various cases and risks of Cybersecurity, and Sect. 8 Types of Cyber-attacks. Finally, in Sect. 9, the chapter is ended with a conclusion.

2 Smart City

There has been a dramatic gradual growth in the population density of the world, especially within the metropolis which requires an advancement in all aspects of a citizen's life. In general a proper built-up city includes several complex structures like human actions and manners, infrastructure, technology, formation of communal-legislative, and the economy [5]. A smart city opens up ways to intelligently supervise various appliances such as people, environment, health system, education system, transport system, energy grids and resources, as well as homes and apartments, etc. The architecture of a smart city comprises of ICT which includes some of the prominent hot topics of the current technological affairs, like IoT, Cloud Computing, Big Data, etc. [5]. The concept of the smart city assists the globe to evolve and expand, spread and facilitate sustainable progress to mitigate the threats and meet the demands of the rising globalization. Restructuring the city by means of promoting innovation, introducing strong and intelligent administration, and ensuring resilience and sustainability has been the key aspects of establishing the smart city.

This development contributes towards a better, efficient, and interactive environment in order to manage, monitor and control all technological services smartly shared and connected within the applicable nodes of a smart city. As well as the data and information are exchanged conveniently making it easier to take decisions based on environment requirements. Smart City [6] is a paradigm to conceptualize both tangible and behavioral practices of a city that includes citizens, services provided to them and the substantial infrastructure to the digitized territory by means of interoperability between technical components such as sensors, actuators and their abilities to process the gathered data.

3 Opportunities of a Smart City

The prime pinnacle of the notion of the smart cities is the rising of the standard and coming up with efficient solutions for various services to be provided in the metropolitan area. Internet of Things (IoT) is one of the paradigms that is best-known for its ability to build and control as well as monitor services of smart applications for intelligent society [7]. IoT works by integrating sensors with every object of the surroundings and assists to interconnect them by sharing data so as to convey many services to the users and this is generally done by means of communication over the network/internet.

As stated above IoT uses sensors but along with this, it also utilizes technical nodes such as electronics and software to control and manage how all the parts of the system/application works. Every single object of the system generates and captures data from the environment and passes them on via sensors to other objects of the application. IoT and Cloud Computing technology are two different ideologies but plays a key role in the smart life based on internet as their absolute adoption in our regular lifestyle is expected to give rise to popularity resulting in making them the significant elements for the future internet. On the other hand, Cloud Computing allows to integrate data and information from various data sources on request over internet by giving out computing services to the network nodes. Both Cloud Computing and IoT has its merits and demerit with several complications and challenges to be faced by the application users in the near future where the smart network will absolutely rely on them for a better and effective solution to all the hassles of a general application. In the last few years, IoT has been seen to gain the popularity by rapidly evolving and grabbing the attention of the technology-based inventions to lead an easy and convenient life.

4 The Prominent Characteristics of a Smart City

Smart City is utilised to build a city with intelligence where one is capable of manufacturing items, educating residents and govern involving new technological aspects in everyday life.

A city can be considered as competent and efficient if and only if it has the following features [4]:

i. It is significant to be able to recognize and analyze urban mobility as an important resource. Urban mobility is a crucial element of a smart city as it majorly contributes towards its economy and promotes a smart and resilient evolution of the city. It is expected that in near future several innovations in urban mobility will eventually result in a scenario where the citizens and consumers will be able to have the benefit of on-demand smart mobility options in an affordable range with a replacement of the currently provided transport services to all for the same purpose.

ii. At present the digitised era is all set to offer a number of job and business opportunities to the people. An intelligent city needs a set of intelligent workers who
can uniquely devote themselves towards building a smart city which will in due
course return to them as a strong and intelligent economic factor. The advancement in technologies are able to simplify the procedural works of government
and give out smooth and effective experience to the companies. A smart city can
play a major role in offering new business opportunities to the citizens making it
more engaging and options available to show how an intelligent economy works
for the people and by the people.

iii. Providing smart education system for people of all ages, races and religions
will play an important role to bring a lifetime change in the everyday life of the
residents. People staying in a smart city has to concentrate on various educational styles which are intelligent and enables a number of job opportunities
and choices for everyone. If we want to see a legit change and rise in economy,
offering intelligent and smart education system has to be a key point to reach
that certain goal.

iv. City residents opt towards building lifestyle based on technological solutions
by correlating all the aspects of the available technical components. This way
all the activities performed at a regular basis contributes towards providing a
innovative, faster, reasonable and a simpler and sustainable life learning and
living opportunities for the citizens. A smart construction system aims to offer
intelligent building systems which are interconnected via technical components
bringing in a number of features so as to give rise to the level of efficacy,
protection and ease of the population. The IoT paradigm combines with the
traditional building/ construction system in order to offer automated systems
to gather and analyze necessary data for surveillance and management of the
application when and where needed on demand.

v. Healthy green city has always been at the core of the conceptualisation of a smart
city in order to enhance the biodiversity of a city. A smart city must have its
attention on bringing vital changes to improve the green energy tools and keep
the nature intact and fresh for a healthy lifestyle inside the city. The enhancement
of the waterways, green life, and offering sound and sustainable environment to
the inhabitants are the fundamentals of a smart city which should always remain
as a prime focus to act in accordance with the idea of sustainability in a city life.

vi. There are certain stakeholders involved in a smart city plan where they play a
key role in governance. Traditional method of governance has to be improved
by means of introducing intelligent services such as appliances to handle several
schemes and programs needed for the development of the legislative structures
and enhance court decisions. Along with these, to be able to enhance the activities performed in order to increase the standard of governance provided to the
citizens by delivering a vital number of services and goods to facilitate regular
practises as a part of innovation. The smart city needs smart plans and strategies, technical aspects as well as several crucial business models and a number

of smart technological components, intelligent people, proper laws and procedures, and sustainable knowledge to improve the efficiency of governance in
it.

5 Applications of a Smart City

To acquire sustainability as demanded by all the involved elements of a smart city
and to give rise to the standards of living, a number of decision makers utilises a
neighborhood paradigm and technologies relating to big data to predict outcomes.
Several technical solutions as aforementioned are practised in the smart city to ensure
an effective and innovative communication within the stakeholders enabling a good
lifestyle in order to form a strong smart urban infrastructure. This results in offering
a well-established communication within the connected nodes with appropriate data
analysis in a reasonable range making it a cost-effective solution. Data mining is a key
element for digitisation and a proper exploitation of gathered data works wonderfully
for several enterprises and business utilities which also involves a domain of smart
city.

The generated data is huge in number and cloud supply is extremely a beneficial
choice as an efficient storage which is gradually changing into an up-to-date version.
With the exponential growth of IoT in every industry, this integration of cloud and
IoT might successfully bring a vital change in the technological-based society. The
following are the applications [4] of the smart city which are prominent and can
elevate and ease the lifestyle of the inhabitants as stated in Fig. 1.

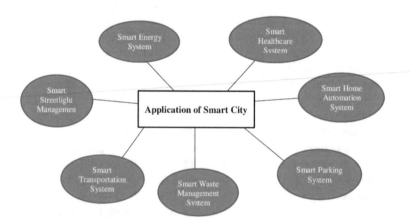

Fig. 1 Applications of Smart City

5.1 Smart Healthcare System

An amalgamation of technologies utilised in the healthcare sector with the motive to maximize the quality of healthcare to the patients and create a convenient ambience for the healthcare workers by allowing them to stay connected to the gathered data at any time of the day makes it a bit easier for both the parties to enhance the quality of life. The use of intelligent components and smart devices in this sector can highly assist in providing accurate diagnosis and near to perfect healthcare services to the involved parties which can bring a revolutionary change by increasing the survival rate of the patients [8].

5.2 Smart Energy System

It is particularly known as an approach taken to achieve a prime solution for gas grids, thermal and electrical power system. Smart energy system comprises of digital network which contributes in offering clean and green energies for smart cities. To an extension smart solar system or plants for wind power might play a significant role in the ecosystem of smart city [9].

5.3 Smart Transportation System

Smart mobility refers to a clever network of transport and portability having inter-connected nodes of technical elements involved in building a smart infrastructure of transportation system for regular use in basic life, enterprises and businesses life [10]. This application incorporates several technologies and updated services in order to improve the overall experience of the transportation for the citizens and regulate traffic issues to ultimately avoid and mitigate injuries of the inhabitants [11]. The advancement of the sizes and trends of the available transports in the metropolitan has given rise to particular problems that exert influence on the welfare and protection of the city. An analysis portrays that urban transports such as buses and trains have always been at the position of dominance over regular private transports despite of them being able to satisfy the basic needs of the inhabitants like privacy, independence, flexibility, etc. On the contrary, these transports at times tend to have a negative impact on the atmosphere giving rise to climate issues as they make extensive use of conveyance which originates noise and air pollution [12].

5.4 Smart Parking System (SPS)

SPS is referred as a solution to the arising problems of urban mobility. This application with the combination of technologies and its vital components enables to gather real-time data about the availability of parking and grants the inhabitants to know about the updates on the traffic congestion and conditions of roads. One of the major concerns of a busy city and its inhabitants is to look for parking areas and the technology behind smart parking system predominantly supports the people/drivers to find out spaces for parking mitigating their confusions and assisting them to manage time for the same. Now-a-days many companies are working on providing these services to the urban people so as to ease their life for good and majorly contributing towards an easy lifestyle in a busy city [13].

5.5 Smart Waste Management System

Smart applications such as waste management has been introduced which deals with discovering the status of the litter basket. The improper settlement of the left-over foods and waste products in the cities have been a burning issue as these spreads smell around the public environment. Being able to notify the inhabitants about the updates of the waste bin (i.e. smell level and waste) assists them to manage it properly on time saving the ambience from the unnecessary accumulation of dirt and foul-smelling.

Another similar solution to collect garbage has been proposed which effectively works on alerting the city authority about the time and location of the waste that needs to be collected. This aids in proper management of the waste-collection ensuring a healthy ecosystem for the inhabitants [14].

5.6 Several Other Related Applications of Smart City

In order to put an end to energy waste, a system for streetlight has been introduced which primarily works depending on the base server that relies on transmitting the gathered weather data. This helps in reducing a number of injuries and accidents saving the lives of hundreds of the citizens.

There are people who have the habit of gardening and to offer them the service of smart gardening can greatly assist them to keep a note on their regular readings of the environmental conditions. Kumar et al. [15] has proposed such system which aims to recommend the application user about the updates of every concerned matter and enabling him/her to assure the organic yield is safe and remains intact throughout the season.

A home automation system [16] has been proposed where it provides the members of the house to enjoy the benefits of having an intelligent household. The application allows the users to get notified about their activities related to the house such as keeping periodical notes of the associated bills and gas consumption as well as tracks and manages the home appliances remotely. It has been observed that the clever and smart household ensures in providing a safer, reasonable and energy-efficient home services to the users.

Tracking climate parameters and the environment around one has been a significant factor for a better staying and living a good healthy life specially if it is cost-effective and saves energy with a good purpose. Parmar et al. [17] has proposed to intelligently track and keep records of the weather and climate conditions. It eventually enables the application users to select the correct and appropriate environment-friendly place to live, facilitating the overall health conditions for them.

There are several other similar services which are available and are user-friendly making them the best fit in order to provide a well-balanced, healthier, and a safer lifestyle to the urban residents granting them to take the real essence of staying in a smart city surrounded by a well-managed and controlled technological components.

6 Cybersecurity and Smart City

With the rapid increase of pollution and congestion as well as the demand to keep a balance in energy and sanitation services, it has become a prime necessity to take good care of the regular lifestyle of every resident. A city has to have the capability to control and absolutely maintain the sustainability of the vital changes that are occurring around it. And thus introducing smart solution has been the prime focus of digitised services provided to the inhabitants. As of now, a good number of cities have already adopted smart solution to lead and offer a better way of life to their residents which is primarily known as a "Smart City" as aforementioned in the earlier paragraphs. Adopting such a model for a city allows it to render enhanced services relating to administration and operational procedures with the motive to lead towards a maintainable environment for the natives. Smart solutions throughout the evolution of technology have been laying out which has given rise to the a rapid economic growth and supported the business aspects. Smart projects and applications have been widely contributing towards achieving the sustainability development goals (SDG) which leads to evolving a shielded smart city but are at a higher risk of facing cyber threats and cyber-security attacks [11].

An analysis on the various security events by Forbes portrays that there has been a drastic increase in cyber-attacks on smart applications in the recent years [18]. As per Hassija, the evolution of smart solutions has essentially given rise to security and privacy issues. Few of the expected cyber-attacks that can take place due to the applications of smart city are as follows [19]:

i. It is possible for any intruder to harm the city by manipulating the traffic lights as they are now prone to cyber-attacks due to being available over wireless medium.

ii. Cyber-intruders can mislead the inhabitants by infusing wrong details of several routes while rendering smart transport services to them.

iii. The cyber-attackers can avail the opportunity of disintegrating the power grid resulting in permanent black-out for a longer period of time.

iv. The water-supplies can crumble due to modification of their chemical balance in an inappropriate way leading to vital public health issues.

v. The intruders/hackers can easily keep an eye on the movement of the inhabitants and cause harm to them in the long run by means of spying them through surveillance cameras which gives the hackers a direct access to their personal and private details which are crucial.

7 Various Cases and Risks of Cybersecurity

The reports says that in the year 2015, the world has witnessed a power outage due to the cyber-attack that took place in one of the cities of Ukraine which caused electricity failure for almost around an hour for the people residing there. Similar scenario was seen in 2019 but in the United States of America where the computing devices of the governments were attacked and were blackmailed to give out 13 bitcoins in return of the encrypted files in them [7].

If the city somehow fails to take the control of the smart applications and their services offered to the residents, it can in the end lead to a disastrous technological outcome where the city might be forced to witness several circumstances of economic failure or deteriorated quality of services putting everyone's lives at risk. The incorporation of ICT in the daily life has not only given rise to merits but also as some associated security and privacy risks which cannot be ignored and the administrative authorities are bound to take necessary measures against them. In some cases, ensuring cyber-security can be one of the major constraints of the services provided by means of initiating a smart city. In such a framework, it has become a prerequisite for the city planners to consider and propose various mechanisms to combat cyber-attacks by presenting and adapting new guidelines, updated policies and control appliances for the same. The six crucial standards of any smart city are the social implications, emotional and analytical intelligence, maintaining the excellence and best practises, environmental balance, as well as the level of innovation and creativity. The proposed paradigm in the paper assists by incorporating several techniques based on cognitive security that easily contributes in assessing the risk levels of the cyber-security in proportion to gathering and handling large amount of data in the smart system environment.

As smart application is all about having inter-connection between several technologies and certain protocols, there always remain certain challenges that are encountered due to its diverseness. This in due course escalates the security threats and risks of cyber-security. The cyber-intruders look for ways they can execute attacks over gathered data and they tend to do these via targeting the vulnerabilities of the smart system. One of the papers by [18] suggests that there are five fundamental layers comprising of network, operating system, hardware, software and firmare in the set up of the Internet of Things which makes it more prone to vulnerabilities. There is a higher possibility of being able to disrupt the smooth procedure of smart systems due to various petty issues such as easy passwords, a unencrypted traffic as well as via default configurations which are now easily attainable through tools for instance SHODAN giving rise to the number of cyber-attacks.

Furthermore, as mentioned by English et al. [20] the attackers can also misuse the weak default passwords to get into the system via constructing attacks related to memory buffer. Such attacks can upsurge the entire systematic behaviour of the application and create ruckus in it by manipulating the technical components. Another study by Mishra and Dixit [21] recommends that if a device is a part of any mesh network, it is possible that the intruder could absolutely derange the complete confidentiality of the network. The study of Benkhala et al. [22] have raised the point that it is possible for the attackers to utilise the spoofing attack, i.e. the main issue here is that the recognition of the components are usually susceptible and transferred during the registration of the devices in the server. There is another chance of modification of the transmitted messages as well as can also damage the data transmission via flipping attacks in the network. In a narrative of Ling et al. [23] it has been portrayed that in a smart plug system there are four kinds of plausible attacks to ascertain the complete system such as scanning the devices, spoofing and brute force attack as well as the firmware attack.

As it has been already mentioned that ensuring cybersecurity is considered as one of the key challenges of smart city and as per Ijaz et al. [24] there are 3 certain factors that highly affects the privacy and security of a smart city. At first it is the technical component which involves IoT, cloud computing, artificial intelligence, databases, software and the semantic web. Along with this, the second component is the governance which comprises of various domains of a smart city like health, education and environment. Last but not the least, the third factor is the socio-economic which directly deals with the business enterprises, banks and finances, etc. Despite of technology playing a key role in developing a smart city, it has it pitfalls on the same line. Smart applications like power grid management, biometrics, phones, etc. have their concerned security and privacy issues. The cyber-attacks that are held on the smart systems are versatile and several kinds of pathways to attack these systems are viable due to availability of different technical components as a part of the smart city solutions. It is crucial to form and install strategies against the cyber threats and attacks to ensure data security. The first and foremost strategic stage that a smart city can acquire is to facilitate the situational recognition of the robustness and fragility of the smart system to combat cyber-security. In addition to it, since smart city is all about integrating technological components to enable widespread sharing of data and information over the network, thus it is required to take steps for establishment of appropriate measures to create awareness in favour and against of the technological aspects utilised for developing the system.

8 Types of Cyber-Attacks

According to Alkeem et al. [19] and Liu et al. [18] there are few common expected kinds of attacks which are faced by the IoT or network of smart and connected nodes as given in Fig. 2. These are as follows:

a. Eavesdropping enables the intruders to block data transmission and assists them to acquire subtle facts and information.
b. Data modification allows the hackers to bring in unnecessary and harmful changes to the data collected from the network and efficiently stored on the server.
c. The attackers using replay attacks eventually creates confusion by delaying the original message sent to the receiver after a short period of time.
d. The denial of service (DoS) attack basically deals with initiating traffic congestion by flooding the server with unnecessary contents which is even out of the holding capacity of the system.
e. The attacker involved in the Man-in-the-middle attack positions him in between the two involved parties to interrupt the basic data communication.

There have been several circumstances where such type of aforementioned attacks have taken place in the smart city applications such as on its prominent and mostly used technological components like Radio-frequency identification (RFID) and Near-field communication (NFC), Wireless Sensor Network (WSN), Machine to Machine (M2M), smart grid, and smartphones.

However, it is basically believed that the IoT framework comprises of three layers namely perception, network and application but they do not have any exact standardized names. It is significant for the ruling authority of the city to assess the threats of cybersecurity in order to offer data privacy and assure quality of life for the inhabitants living in a smart city to enhance the paradigms. They must consider and record the necessary and fundamental requirements of the architecture before installing any smart component on the network and possibly ignore any sort of default configuration over the network. They should also be able to prevent the services of the city from various attacks like DoS or data theft which highly falls into the major category of security breaching. The acamedics should always be in touch with the state of

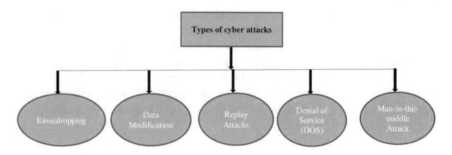

Fig. 2 Most common types of cyber attacks

the art of the evolving cyber-attacks which in the end will predominantly contribute in enhancing the overall cybersecurity scenario by granting the city managers the solution to prevent the smart city appliances from threats and risks of intrusion and security breaches. The legislators must procure the needful information that enables establishment of laws and policies to reinforce security and privacy in the smart city. It is significant to keep a note that while evaluating a city's security might definitely vary from assessment of any other city of a nation, every city has its unique dynamic. Strategies practised in a city might not work well in the other so it is essential to accept and understand the viewpoint that same solution might work differently for two separate problems of the city with a variation in their levels of security. The paper has proposed a solution to deal with the above-mentioned issue in the context of smart city, i.e. a model for measuring relative security can be developed on the basis of the surface attacks in a city. They have identified and suggested four measures, such as utilising the group of components used in the system to establish an attack surface, recognising the attack vectors that enables attacks on the formed surface, generating strategies to combat these attacks by mitigating the surface which lastly leads to assess the relative security by developing enhancements in the security and privacy. Basically the level of threats in the Cybersecurity could be measured and regulated based on the various affects of the attacks held on socio-ecological and economic factors on the smart city stakeholders.

9 Conclusions

Since the last decade, the society has been hearing that the integration of the smart technologies could bring in a rapid change in their lifestyle. And now coming to reality, it has been perceived that the back-to-back innovations and technological evolution in the recent years have greatly influenced the everyday life of every citizen around the globe. Smart City is one of the most favoured concepts in today's world and every nation is trying its utmost to adopt it for its growth and expansion. The pertinent utilisation of ICT is mandatory and invaluable for ensuring and dealing with an intelligent and smart city. The main aim of this chapter is to focus and highlight on the several aspects of the smart city as well as how and why cyber-security is much needed to assure the genuine utilisation and mitigate challenges faced by the implementation of a smart city in a nation. As the infusion of both is in its early stage, thus there are various factors which have strong influence over the spread of this technology. The objective of this chapter is to introduce those concepts to the readers and give a thorough idea on how the correct and suitable usage of both factors can widely change the societal perspectives towards life.

References

1. Jeong Y-S, Park JH (2019) Iot and smart city technology: challenges, opportunities, and solutions. J Inf Proc Syst 15(2):233–238
2. Abdulrahman LM, Zeebaree S, Kak SF, Sadeeq M, Adel A, Salim BW, Sharif KH (2021) A state of art for smart gateways issues and modification. Asian J Res Comput Sci 1–13
3. Sadeeq MM, Abdulkareem NM, Zeebaree SR, Ahmed DM, Sami AS, Zebari RR (2021) Iot and cloud computing issues, challenges and opportunities: a review. Qubahan Acad J 1(2):1–7
4. Hassan RJ, Zeebaree S, Ameen SY, Kak SF, Sadeeq M, Ageed ZS, Adel A, Salih AA (2021) State of art survey for iot effects on smart city technology: challenges, opportunities, and solutions. Asian J Res Comput Sci 22:32–48
5. Abdulqadir HR, Zeebaree SR, Shukur HM, Sadeeq MM, Salim BW, Salih AA, Kak SF (2021) A study of moving from cloud computing to fog computing. Qubahan Acad J 1(2):60–70
6. Washburn D, Sindhu U, Balaouras S, Dines RA, Hayes N, Nelson LE (2009) Helping cios understand "smart city" initiatives. Growth 17(2):1–17
7. Andrade RO, Yoo SG, Tello-Oquendo L, Ortiz-Garcés I (2020) A comprehensive study of the iot cybersecurity in smart cities. IEEE Access 8, 228 922–228 941
8. Georlette V, Moeyaert V, Bette S, Point N (2020) Visible light communication challenges in the frame of smart cities. In: 2020 22nd international conference on transparent optical networks (ICTON). IEEE, 2020, pp. 1–4.
9. Pawar L, Bajaj R, Singh J, Yadav V (2019) Smart city iot: smart architectural solution for networking, congestion and heterogeneity. In: 2019 international conference on intelligent computing and control systems (ICCS). IEEE, pp 124–129
10. Mustafa YT, Sadkhan S, Zebari S, Jacksi K, Recent researches in earth and environmental sciences
11. Arroub A, Zahi B, Sabir E, Sadik M (2016) A literature review on smartcities: paradigms, opportunities and open problems. In: 2016 international conference on wireless networks and mobile communications (WINCOM). IEEE, pp 180–186
12. Razaq HHA, Gaser AS, Mohammed MA, Yassen ET, Mostafad SA, Zeebaree SR, Ibrahim DA, Abd Ghania MK, Farhan RN (2019) Designing and implementing an arabic programming language for teaching pupils. J Southwest Jiaotong Univ 54(3)
13. Srinivas M, Benedict S, Sunny BC (2019) Iot cloud based smart bin for connected smart cities-a product design approach. In: 2019 10th international conference on computing, communication and networking technologies (ICCCNT). IEEE, pp 1–5
14. Chen W-E, Wang Y-H, Huang P-C, Huang Y-Y, Tsai M-Y (2018) A smart iot system for waste management. In: 2018 1st international cognitive cities conference (IC3). IEEE, pp 202–203
15. Kumar Y, Rufus E (2018) Smart kitchen garden using "biothrough" at a low cost. In: 2018 fourteenth international conference on information processing (ICINPRO). IEEE, pp 1–3
16. Singh H, Pallagani V, Khandelwal V, Venkanna U (2018) Iot based smart home automation system using sensor node. In: 2018 4th international conference on recent advances in information technology (RAIT). IEEE, pp 1–5
17. Parmar J, Nagda T, Palav P, Lopes H (2018) Iot based weather intel-ligence. In: 2018 international conference on smart city and emerging technology (ICSCET). IEEE, pp 1–4
18. Liu X, Qian C, Hatcher WG, Xu H, Liao W, Yu W (2019) Secure internet of things (iot)-based smart-world critical infrastructures: survey, case study and research opportunities. IEEE Access 7:79 523–79 544
19. Al Alkeem E, Yeun CY, Zemerly MJ (2015) Security and privacy framework for ubiquitous healthcare iot devices. In: 2015 10th international conference for internet technology and secured transactions (ICITST). IEEE, pp 70–75
20. English KV, Obaidat I, Sridhar M (2019) Exploiting memory corruption vulnerabilities in connman for iot devices. In: 2019 49th annual IEEE/IFIP international conference on dependable systems and networks (DSN). IEEE, pp 247–255

21. Mishra A, Dixit A (2018) Resolving threats in iot: id spoofing to DDoS. In: 2018 9th international conference on computing, communication and networking technologies (ICCCNT). IEEE, pp 1–7
22. Benkahla N, Belgacem B, Frikha M (2018) Security analysis in enhanced LoRaWAN duty cycle. In: 2018 seventh international conference on communications and networking (ComNet). IEEE, pp 1–7
23. Ling Z, Luo J, Xu Y, Gao C, Wu K, Fu X (2017) Security vulnerabilities of internet of things: a case study of the smart plug system. IEEE Internet Things J 4(6):1899–1909
24. Ijaz S, Shah MA, Khan A, Ahmed M (2016) Smart cities: a survey on security concerns. Int J Adv Comput Sci Appl 7(2)

Cyberbullying Instilled in Social Media

Moinuddin Zubair, Shahrin Zubair, and Mohiuddin Ahmed

Abstract In the everchanging world, technology holds up a very big space and is presently infused into almost every sector. And a vast portion of this technological world is influenced by social media. This very impact of social media is not limited within the current boundary of its domination rather it's expanding with each passing day. With this growing influence, although it's playing a significant role into bringing this world in a more closely defined arena, it is also attracting a huge number of illicit activities in the name of cyberbullying and scamming. Holding up with this trend, people are sharing more and more of their personal details which includes personal identity, their bank accounts a lot more. In a nutshell, social media is like an open book into the life of a person which does leave some loopholes open for the people to take advantage and use it in their favor. Accessing this book, these cyber-criminals get to use the precious information and use it for blackmailing and bullying only for the purpose of fulfilling their illicit demands which puts the targeted person on a loosely constructed pedestal. As per statistics, it is responsible for the loss of hundreds and thousands of teenagers every year as they suffer from insecurity and stress due to the fear of their personal or private data being leaked. Machine Intelligence plays a key part in mitigating the concept of cyberbullying in social media with respect to personal details and statements given by the users. Several machine learning techniques have been utilized and researched for the same and this chapter focuses on highlighting all the procedures currently used to limit cyberbullying in this area.

M. Zubair (✉)
Department of Computer Science and Engineering, Brac University, Dhaka, Bangladesh
e-mail: moinuddinzubair26@gmail.com

S. Zubair
Student Connect BD, Chattogram, Bangladesh

M. Ahmed
School of Science, Edith Cowan University, Perth, Australia

1 Introduction

Social Media is the at the pinnacle of the success achieved in the field of technology and science. Connecting the people miles apart from across the world in one single platform that allows the user base to share information in the visual, text and audio format is no less than a fiction if narrated to someone from the 1900s. Every success has its own share of downsides as well and so is the scenario with the evolving technological based world. Cyberbullying is at a constant rise and it is at an uphill with the passing of time and the evolution of this virtual mini world [1]. Several hundreds of papers have already been published illustrating the role of social media as a tool of making the lives of people across the world easier, simpler and richer and side by side highlighting its fair share of drawbacks. And when the topic of concern is social media, the keyword "Cyberbullying" is also not alien to us. Although this paper is also going to shed light on the usage and snags of social media, its main pivot would be to exemplify how machine learning has been a key tool in confronting cyberbullying by a large margin and how crucially it has been used utilized in bringing down the iniquitous and wicked cases of bullying happening online alongside unveiling the monstrous personalities behind the scenes [2]. At this stage it is very crucial to pinpoint the issue of cyberbullying and the ways of dealing as with the onset of the COVID-19 with the rise of covid cases across the world, there has been a sheer peek rises in cases of cyberbullying across the different social media platforms. Just after the restrictions implemented during the COVID-19 period, cyberbullying has increased by 70% in just a few months, according to the L1ght, a group that monitors online harassment. According to recent studies, while school bullying drastically decreased during the COVID-19 pandemic, cyberbullying remained quite steady. Thus, Cyberbullying is a phenomenon which is going in hands to hands with the evolution of technology and which is increasing every then and now with the rise of greater dependency on social media platforms. Mentionable is that although with the advent of technology, criminal cases based on online increased drastically, several tools to counter these cases of harassments, crime and threat also showed up which did wonders in attenuating cyberbullying and striking a fear across the hearts of the bullies sitting behind their computers with a mask on. Machine Learning has helped dealing with this huge influx of cyberbullying by creating models of algorithm which assists the developers in filtering out the troll profiles, predicting future cyberattacks, locate and spot the bully and bringing the victim to justice.

This chapter is assembled in an order where Sect. 2 describes what Cyberbullying is followed by Sect. 3 which talks about the influx of Cyberbullying mainly during the pandemic. Section 4 provides an introduction to who these bullies are and how ado they end up indulging themselves in cyberbullying. Section 5 states regarding the impact of cyberbullying on the victims and how their life gets affected due to it. Section 6 is divided into 3 sub branches where we introduce machine learning, talk about the types of machine learning in Sect. 6.1, followed by the impact and use of machine learning in Sect. 6.2 eventually followed by how machine learning models are assisting in combating cyberbullying in Sect. 6.3.

2   Cyberbullying

Bullying that is committed across online platforms is known as cyberbullying. Social media, services based on texting and gaming platforms are all potential venues for cyberbullying to take place. It is a pattern of behavior mainly targeted to bully, terrify, mortify and infuriate the victims solely earmarked on the basis of selfish personal reasons. This sort of bullying sometimes coexists both in public and online as well. However, it is mentionable that cyberbullying does leaves a digital trail behind which is basically a record that can be informative, handy and may offer practical proof to put an end to the abuse [3]. It has been a while since cyberbullying has been a constant part of the virtual world on which teenagers spend most of their time on. On whichever platforms there is a presence of even a mild amount of audience, cyberbullying makes it base and puts a footnote of its trace be it gaming industry, social media platforms or different blog sites. Online harassment doesn't only include threats, abuse and blackmails committed over texts, but it also includes using a false identity to send salacious messages to others on that person's behalf, body shaming next to the comment sections of the pictures posted online etc. [3]. Even though cyberbullying has been prevalent for a long while, the pandemic has made the conditions worse. Barreto attested that teens' greater reach of access to the Internet along with newer electronic gadgets and feasibly lax online activity monitoring are all contributing factors contributing to the rise in cyberbullying. She also concluded, "Within the discipline of psychology, the study has varied on its viewpoint on the link and interaction between cyberbullying and traditional bullying. Because cyberbullies can remain anonymous, bullying can happen at any time. This makes cyberbullying a potentially greater and more dangerous kind of hostility than the regular traditional bullying [4].

3   Upsurge in Cyberbullying

From the start of the pandemic in the year of 2020, as the classes conducted in the educational institutions shifted online, the rise in the cases of cyber bullying exponentially surged. As the daily activities which were on a perpetual basis regulated offline, a sudden shift in their administration being done online set a whole new path for bullying to make its own new pavement. Not only educational affairs but the civil, commercial, healthcare activities also made their new base on the virtual peripheral. And in the recent scenario of sudden shift online, it was observed that the students who are prone to bullying are more likely to indulge themselves in cyber bullying. A rigorous strenuous torment on these teenagers enrages them to seek out vengeance motivated by their constant need of approval which they fail to receive in person. So, rather they choose to sit in front of computers wearing mask hidden from the world with an aim to look for reprisal as a retort to the constant bullying they received which goes on feeding their due ego. As per school counselor

Anna Weddington, the pandemic along with the increased remote activities has made more room for an increase in cyberbullying cases. Due to the movement of social groups from the offline to the virtual platform, the students sometimes come across communicating with people who are total strangers to them who they don't see in person on a daily basis nor they have any picture to prove the identity or the camera on. So these social media platforms opens a pathway for these bullies to get access to certain pictures of the users posted online and some website links of which these bullies make an illicit use and even before the victims come to contemplate the situation of what's going to happen, they already get themselves prone to dangerous circumstances [5].

Speaking about the virtual platforms, Zoom has already come under the hit list for the neoteric disruption of random strangers who went on hacking virtual business meetings with indecent and despicable words. A record number of reports were received by the New York City DOE that documented the abuse and security breach across the Zoom platform by innumerable clients which resulted in DOE taking away the permission for the usage of Zoom. Boredom during the pandemic situation while residing at home also fueled up many teenagers to engage themselves in some sort of bullying. They took this up as a sort of entertainment. There were also reports as per the parents of Asian American children who disclosed instances of their kids being mistreated on the basis of their skin color and race and looking down upon the Chinese kids referring to the origin of virus to China calling it as "The China Virus" or "The Asian Virus". Attacks on them also surged as the news of the virus being called the China virus circulated online. The more this sort of news circulated, the more Asian kids became victim to the new attacks. There was also an efflux of tweets and posts targeting Asian kids demanding a ban upon them which affected the emotional and mental health scenario of these Asian kids sidelining them to long lasting depression [6].

According to Center for Disease Control (2017), 20.2% of the high school students are bullied on school property and a whopping 15.5% of them are the victims of cyberbullying [7]. The percentage of individuals experiencing cyberbullying at some point of time in their lifetime have almost doubled from 18 to 34% from the period of 2007–2016 which is on a constant increase as the years go by [2].

4   Who Are These Bullies and What Causes Them to Get into Cyberbullying?

- **Increased distress**: Increased schoolwork pressure, social pressure, family pressure stresses some teenagers up and to release their stress, they end up indulging themselves in activities ending up harming others.
- **Isolation and Loneliness**: Often times, many teenagers suffer from problems of isolation and depression. Some do not feel close to family members at home, others have deteriorating bonds within close friends' group and some have dating

issues. To cope up with these, these teenagers end up lashing out their anger by getting themselves involved in cyberbullying.

- **Not enough supervision**: A number of parents are not able to look after their kids due to their increased work pressure. So, these kids not getting the due attention from their guardian end up wallowing in these activities out of curiosity and improper supervision.
- **Ignorance**: Some of the teenagers feel outcasted and isolated from certain friend groups which motivates them to seek revenge through the means of threatening and frightening other people online.
- **Being themselves bullied**: Number of cyberbullies are themselves victim of intense bullying in school premises. Not being able to stand against them publicly, they sort out other ways to seek vengeance and to malign and disparage the bullies so they look up after social media platforms in order to frighten the bullies thus earning the title of cyberbullies themselves.
- **Languor**: Boredom and a monotonous routine is one of another factors which motivates some teenagers to yield to their temptations and curiosity of exploring and scaring off other people. They do it out of entertainment purpose not knowing its far reaching effects on the other side of the computer.

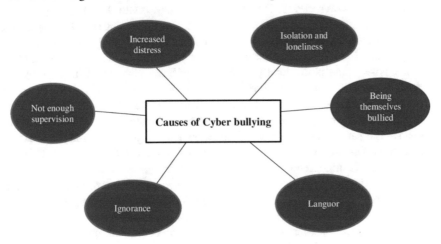

5   Impact of Cyberbullying

With time, Cyberbullying has become an international health concern. Profusion of correlational studies have shown that there is a direct connection of cyberbullying with the mental and emotional health of teenagers. Smallest of the things help teenagers carving their future but in the same mannerism, these things can break them down and push them into an everlasting haunted future. This is the reason why bullying but mainly cyberbullying can be such a threat to the life of these adolescents

[8]. According to Anna Weddington who is a school counselor at Anderson Elementary in New Hanover County, Cyberbullying has some long-term effects which is not only limited to breaking their self esteem but is also linked to altering their overall viewpoint of approaching different relationships and can negatively switch their idea of the world being a safe place for them. She also proclaimed that bullying of this sort creates a certain kind of negative friction among a group of friends and the victims tend to perceive themselves isolated and an outcast. It creates a sense of trepidation and unease for them to visiting school so they end up going to the premise with the horror of being made fun again or being ostracized by those online bullies. This results in their overall grades dropping and affects not only their institutional life but creates a change in their eating, communicating and sleeping patterns as well [5].

6 Machine Learning

Machine Learning is a branch of Artificial Intelligence and Computer Science that mainly anchors on the usage of data and algorithms to predict and imitate the way humans learn with a view to increasing its accuracy. It allows software applications to become more accurate at predicting outcomes without being actually programmed to do so, rather by estimation with the assistance of data and algorithms [9, 10].

Machine Learning, a revolutionary part of Artificial Intelligence is a significant tool whose usage is drastically increasing with the passage of time. It not only predicts the overall trends applying algorithms but rather also gives enterprises and organizations the scope to dig out customer behavior and patterns of operation. It also supports the booming of new products and services targeted on customer interests and similar sort of patterns. Most of the renowned companies around the world be it Facebook, Google, Amazon, Alibaba, Microsoft, Uber, Netflix, Amazon Prime etc. have made Machine Learning a pivot of their main operations and all their activities are centered around the data retrieved through the application of machine learning. It has surged out as a very significant differentiator in case of competition between commercial organizations in terms of extracting profits [9, 10].

The usage of machine learning is frequent, perennial and quotidian in most of the technologies and applications we use. One of the most common examples is Recommendation Engines. Other common uses of MI include spam filtering, fraud detection, business process automation (BPA), malware threat detection etc.

Mostly, technological advancements based out of storage and processing power have advanced the surge of smart tools carried out solely on the dependency of Machine Learning. Netflix's recommendation engine is one example associated to it. Other than that, Tesla, the electric car company of the Richest person in the world, Elon Musk derives its self-driving capability solely by its dependency on statistical methods and algorithms. These statistical methods and algorithms help uncovering key insights which aids in recognizing patterns [9, 10].

6.1   Types of Machine Learning

Machine Learning is mainly categorized on the basis of how an algorithm works, approaches on recognizing patterns and making predictions. It is classified into 4 categories namely supervised learning, unsupervised learning, semi-supervised learning and reinforcement learning. The type of algorithm which have to be used is chosen by the data scientists based on the kind of data they want to predict [9].

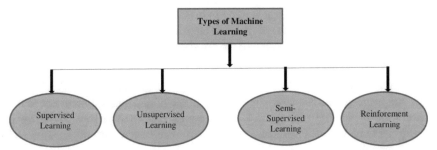

Supervised Learning: The type of data in which both the input and output is specified is known as Supervised learning. In this category of machine learning, the algorithm is supplied with labeled data and the variables are also defined. The output is already predetermined in this type of machine learning [9].

Unsupervised learning: The type of data in which the input and output is not clearly specified is known as unsupervised learning. In this scenario, the output is not predetermined and the algorithm is trained on data which is not labeled. Instead, in this type of machine learning the algorithm sorts out and scans through data to look for meaningful connections and patterns [9].

Semi-supervised learning: Semi supervised learning is a type of machine learning which is an amalgamation of both supervised and unsupervised learning. The algorithm can be infused by the data scientists which is mainly labeled training data but in this aspect of machine learning, the model has the freedom to choose and develop its own contemplation of the data itself [9].

Reinforcement learning: Reinforcement learning is a type of machine learning in which the machine has to complete a step-by-step process in order to get the task done obliging to a certain set of rules. The algorithm is programmed to complete a task and it's given a list of positive and negative cues based on which the machine conducts the task. The process which will be followed to accomplish the task is mostly decided by the algorithm itself [9].

6.2 Uses and Impacts of Machine Learning

Knowingly or unknowingly, the use of machine learning technology has been injected into most of the things we do in our everyday life and most of the software applications we use. Starting from the mega mammoth companies to smallest of companies in nook corners of the world, machine learning models has been implemented in every sort of chore to filter out suspicious activities and extract profits by understanding customer interests. Starting from social media platforms, automobile industries, mobile phone industries, gaming industries, transport to commercial services, every industry is making use of machine learning tools in this decade and its operation is only at a peek rise. Some of the common most uses of machine learning are described below:

- **Image Recognition**: The feature of image recognition itself is developed and shape using machine learning. With the usage of machine learning, image recognition technology was made possible which is able to predict output for each pixel in an image. Face lock in cellphones, face scan in airports are some of the examples of it [11].
- **Social Media Sites**: Giant social networking sites starting from Facebook, Instagram, Twitter, Snapchat all of them are making use of machine learning technology for improving search suggestions and recommendations. Various product advertisements are also done predicting the areas of interests of the user, thus carving a better-looking news feed [11].
- **Voice Recognition**: Voice recognition is now part of most of the smart devices being developed be it mobile phones, personal computers, tablets or smart TVs. This technology helps extracting data faster just by supplying voice input. Alexa, Siri etc. all are the fruits of this machine learning technology [11].
- **Making Predictions**: Video predictions on YouTube, product advertisements across newsfeed, movie and series prediction in Netflix and Amazon prime by sorting out a pattern applying algorithms and scanning the due data is part of machine learning as well [11].
- **CCTV and Video Surveillance**: CCTV cameras are across every nook and corner of the world. The technology detecting slight suspicious movements in the activities of a suspect, finding out criminals passing through places and detecting their previous track records matching with the facial muscles of the person is all an accomplishment of machine learning [11].
- **For customer support and extracting profits**: Companies are now making use of machine learning to extract data regarding customer interests and making predictions on their likings and disliking for a specific product are monitored closely using this technology. This helps the companies and organizations to make better suggestions across websites to the customers which is also a plus point for these companies financially [11].
- **Self-Driven Cars**: Self driven cars like Tesla are making use of machine learning to analyze and detect traffic, pedestrians, sharp turns, objects in road, animals etc. for better self-driving experience [11].

- **Streaming Sites**: Streaming sites like Netflix, Amazon Prime make the use of these platforms for the clients swift and easier by taking feedbacks, detecting and contemplating the patterns of choices of each client to suggest better movies and dramas in future [11].
- **Detecting trolls, fraud and spam**: Detecting trolls, fake ids, bullies, spam messages and filtering them out have been easier since the advent of machine learning technology thus bringing down cyberbullying and cyberstalking by bounds [11].

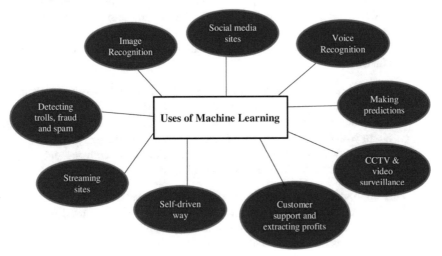

6.3 *Utilization of Machine Learning in Combating Cyberbullying*

The methods stated beneath are varieties of machine learning models implemented to counter and detect cyberbullying under various circumstances. These models of machine learning are already in use across the world and are a key factor in combating cyberbullying and bringing its negative aspects down by a large margin. The methods mentioned here mostly work on the basis of classification where the messages are distinctly judged and classified as a part of cyberbullying or not. The models mainly include the specialized configuration of language features in respect to supervised learning. These are designed incorporating specifically targeted features based on the topics which are known as being prone and common most scenarios to bullying. Rest of the features used are static, social structure features, features which are based on the association rule techniques, sentiment features and vulgar language expansion using string similarity. Machine learning has been applied into different contexts to extract the base concept behind bullying across online platforms. The data available in the virtual world is scanned and the posts, tweets where the victims or other audience talk about their personal experiences are looked after and carefully studied to detect the similar sorts of pattern across platforms for finding out matched traces of cyberbullying. These machine learning models help in scrutinizing and inspecting various psychological issues surrounding the motive and conceptualization regarding bullying. These documentations achieved through the application of machine learning tools contribute in yielding new insights on contemplating the root of cyberbullying and figuring out new features to detect its presence [12–22]

- A proposed method keeps on learning new language indicators which is repetitive and acts as a pattern of bully performed by prevailing bullies across platforms. It is related to query expansion to seek out information retrieval. The whole focus of query expansion is to propound a set of keywords which are correlated and are used on a repeated basis. A number of approaches are used in this regard [23]. For instances,

    - Mahendiran et al. [24] introduced a method solely on the basis of probabilistic soft logic to outgrow a glossary which uses more than one indicators.
    - Massoudi et al. [25] put forward a method which uses temporal information along with co-occurrence with a goal of expanding the query.
    - Another statistical language model was established with an aim of query expansion introduced by Lavrenko et al. [26].

This approach is based on the idea of providing weak indicators of varieties of forms of cyberbullying which includes specific vocabulary used in messages involving cases of bullying. The algorithm then generalizes from these weak indicators to detect possible instances of bullying across the scanned data. The commonest form of vocabulary used by the bullies are discovered by the algorithm which extracts new vocabulary of those involved in bullying. This feedback loop then iterates repeatedly until the algorithm succeeds sorting out a constant score set of scores based on how much the model regards each user to be a bully.

- In tackling the snags of email-based cyberstalking, a model has been developed which includes machine learning, statistical analysis, email forensics and text mining in order to detect and tackle cyberstalking prevailing across emails. This sort of monitoring system framework works on the criteria that it is able to detect and filter cyberstalking mails. It not only verifies and tries to evaluate the issue with the anonymous message but also reevaluates and tries to locate the address of the stalker and extracts necessary information relating to the source. This framework is not only limited to the filtering of these spam mails but rather it enables in manual email evidence collection which assists the authorities to investigate and locate the stalker in order to continue legal proceedings against these bullies [27, 28].
- Of all the social media sites, the platform with the greatest number of cyberbullying cases is Instagram. The most popular social networking site with a large user base only next to Twitter and Facebook faces a tremendous number of issues regarding cyberbullying. Be it spam texts or spam comments under the pictures, the bots and the spam IDs are prevalent on a very large scale. Based on the recent survey published by the anti-bullying charity Ditch the Label reviewed that 42% of the

youth among 10,000 people within a age group of 12–25 found Instagram to be the platform where they were victims of cyberbullying the most. Next comes Facebook at 37% and Twitter at 9%. On a mission to combat this big prevailing phenomenon, Instagram became the first social media platform to make use of machine learning as a tool to eliminate abusive language across its platform. To counter this issue, Instagram announced their strategy of integrating machine learning algorithm to detect the potential cyberbullies, block and eliminate them from the platform. It is making use of DeepText which is the identical machine learning algorithm used by Facebook in the June of 2016 in order to combat their share of cyberbullying concerns. DeepText is a sort of deep learning-based text understanding engine which is able to contemplate thousands of posts per second with near human level accuracy [29]. It makes use of word embeddings to assist the model to understand the distinct way in which the humans make use of language. Its framework is based on functioning like a human brain using deductive reasoning to understand sentences used in varying contexts [29]. Using this algorithm enabled Instagram to cut down its bots, internet trolls with an aim to gain followers or selling their products and other suspected fake ids posting spam comments or hateful texts [29].

- Apart from DeepText, Instagram is also utilizing machine learning for text and image recognition. It is highly assisting the developers of the platform to detect bullying in any sort of mannerism be it video, photo or caption. Instagram also introduced the "bullying filter" to conceal abusive comments. Apart from this making use of the machine learning algorithm the platform can now scan for threats against the user base appearing in posted photos [30].
- Another model of machine learning which is developed and is use in the current world scenario behind which the main idea is that each of all trolling profiles are directly or indirectly followed up by the real profile of the bully hiding behind the trolling one. This estimation is based upon the fact that every troller wants to stay up to date regarding the activities that surrounds his/her fake profile. It is possible to establish a correlation between the characteristic text, structure of sentence making, a repetitive method or idea behind the tweet by utilizing this machine learning technology. Twitter is already making use of this very technology to filter out, differ and seek out the fake profile, correlating this fake profile with the real profile of the user and hunting them down. This technology also scans through the text and content, tries to find existing similarities between the targeted abusive or bully content with the same characteristics shared which are tweeted by the other distinct profile of the user. It is very evident that the troller may change his/her writing style but making co-relations and connecting the dots among the different characteristics shared by the user profile, the followers and the following they share, the tweets they like etc. helps the algorithm to create connections and detect the bully. The preferred time of tweeting or messaging is also another significant point that helps in figuring out the troll and so does help the language and geoposition. Finally, the device through which they usually put out the tweet i.e., their favorite twitter client be it computer, mobile phone or tablet helps create and understand the pattern of the troller [31].

7 Conclusion

Cyberbullying has been a on a constant rise since the advent of technology. And with the influx of the usage of social networking sites among the teenagers, it has been on an uphill. The teenagers are the one who are the most vulnerable to online bullying. These bullies themselves are also teenagers from the same schools and colleges as of their victims and sometimes they are complete strangers unknown to the victim. So, cyberbullying can be specific or it can be randomly carried out just for the fun of teasing or frightening the strangers. Although the new products and models designed and developed utilizing machine learning algorithms are succoring detecting the troll and bully profiles, filtering them out and make these popular social networking sites free of bullies and less hateful, there is still a long way to go. None of the popular social platforms are 100% free of trolls and bullies, rather they are still far from it. But with the pace we are marching towards the future technologies and as new machine learning models are being implemented every passing day, the future is not afar where we will be rewarded with these sites free of any sorts of bullies and abuse. The process is hard and the threshold is very high, but it is quite clear that machine learning has made the job facile and brought down the threshold by a wide margin.

References

1. Micklea Z (17 Mar 2021) Increase in cyberbullying during COVID-19. MIBluesPerspectives. https://www.mibluesperspectives.com/2020/10/12/increase-in-cyberbullying-during-covid-19/. Accessed 15 July 2022
2. Patchin JW (8 July 2022) Bullying during the COVID-19 pandemic. Cyberbullying Research Center. https://cyberbullying.org/bullying-during-the-covid-19-pandemic. Accessed 15 July 2022
3. Cyberbullying: what is it and how to stop it. UNICEF (nd). https://www.unicef.org/end-violence/how-to-stop-cyberbullying. Accessed 15 July 2022
4. Suciu P (10 Dec 2021) Cyberbullying remains rampant on social media. Forbes. https://www.forbes.com/sites/petersuciu/2021/09/29/cyberbullying-remains-rampant-on-social-media/?sh=6c53558f43c6. Accessed 15 July 2022
5. Pandemic blamed for rise in cyberbullying. Cyberbullying on the rise (nd). https://spectrumlocalnews.com/nc/charlotte/education/2021/10/14/cyberbullying-on-the-rise. Accessed 15 July 2022
6. Cyberbullying during COVID-19 (nd). https://www.stompoutbullying.org/blog/Cyberbullying-During-COVID-19. Accessed 15 July 2022
7. Preventing bullying—centers for disease control and prevention (nd). https://www.cdc.gov/violenceprevention/pdf/bullying-factsheet508.pdf. Accessed 15 July 2022
8. Nixon CL (1 Aug 2014) Current perspectives: the impact of cyberbullying on adolescent health. Adolescent health, medicine and therapeutics. https://www.ncbi.nlm.nih.gov/pmc/articles/PMC4126576/. Accessed 5 July 2022
9. Burns E (30 Mar 2021) What is machine learning and why is it important? SearchEnterpriseAI. https://www.techtarget.com/searchenterpriseai/definition/machine-learning-ML. Accessed 15 July 2022
10. By: IBM Cloud Education (nd) What is machine learning? IBM. https://www.ibm.com/cloud/learn/machine-learning. Accessed 15 July 2022

11. Uses of Machine Learning: List of Top 10 uses of machine learning. EDUCBA (2021). https://www.educba.com/uses-of-machine-learning/. Accessed 16 July 2022

12. Dadvar M, de Jong F, Ordelman R, Trieschnigg D (2012) Improved cyberbullying detection using gender information. Dutch-Belgian information retrieval workshop, Feb 2012, pp 23–25

13. Chen Y, Zhou Y, Zhu S, Xu H (2012) Detecting offensive language in social media to protect adolescent online safety. In: International conference on social computing, pp 71–80

14. Dinakar K, Reichart R, Lieberman H (2011) Modeling the detection of textual cyberbullying. ICWSM Workshop on Social Mobile Web

15. Reynolds K, Kontostathis A, Edwards L (2011) Using machine learning to detect cyberbullying. In: International conference on machine learning and applications and workshops (ICMLA), vol 2, pp 241–244

16. Yin D, Xue Z, Hong L, Davison BD, Kontostathis A, Edwards L (2009) Detection of harassment on Web 2.0. Content analysis in the WEB 2.0

17. Nahar V, Li X, Pang C (2013) An effective approach for cyberbullying detection. In: Commun Inf Sci Manag Eng 3(5), 238–247

18. Ptaszynski M, Dybala P, Matsuba T, Masui F, Rzepka R, Araki K (2010) Machine learning and affect analysis against cyber-bullying. In: Linguistic and cognitive approaches to dialog agents symposium, pp 7–16

19. Margono H, Yi X, Raikundalia GK (2014) Mining Indonesian cyberbullying patterns in social networks. In: Proceedings of the Australasian computer science conference, vol 147, Jan 2014

20. Huang Q, Singh VK (2014) Cyber bullying detection using social and textual analysis. In: Proceedings of the international workshop on socially aware multimedia, pp 3–6

21. Bellmore A, Calvin AJ, Xu J-M, Zhu X (2015) The five W's of bullying on Twitter: Who, what, why, where, and when. Comput Hum Behav 44:305–314

22. Ashktorab Z, Vitak J (2016) Designing cyberbullying mitigation and prevention solutions through participatory design with teenagers. In: Proceedings of the CHI Conference on human factors in computing systems, pp 3895–3905

23. Manning CD, Raghavan P, Schutze H (2008) Introduction to information retrieval. Cambridge University Press

24. A. Mahendiran WW, Arredondo J, Huang B, Getoor L, Mares D, Ramakrishnan N (2014) Discovering evolving political vocabulary in social media. In: International conference on behavioral, economic, and socio-cultural computing

25. Massoudi K, Tsagkias M, de Rijke M, Weerkamp W (2011) Incorporating query expansion and quality indicators in searching microblog posts. In: Proceedings of the European conference on advances in information retrieval, vol 15, no 5, pp 362–367, Nov 2011

26. Lavrenko V, Croft WB (2001) Relevance based language models. In: Proceedings of the international ACM SIGIR conference on research and development in information retrieval, pp 120–127

27. Roberts L (2008) Jurisdictional and definitional concerns with computer- mediated interpersonal crimes: an analysis on cyber stalking. Int J Cyber Criminol 2(1):271–285

28. Aggarwal S, Burmester M, Henry P, Kermes L, Mulholland J (2005) Anti cyberstalking: the predator and prey alert (PAPA) system, (2005) in systematic approaches to digital forensic engineering, first international workshop, no. iv. IEEE CPS, 195–205

29. Bayern M, Staff TR, Fernandez R, Okeke F, Miles B, Bohon C, Librescu M (11 Aug 2017) How ai became Instagram's weapon of choice in the War on Cyberbullying. TechRepublic. https://www.techrepublic.com/article/how-ai-became-instagrams-weapon-of-choice-in-the-war-on-cyberbullying/. Accessed 16 July 2022

30. BBC (nd) Can this technology put an end to bullying? BBC Future. https://www.bbc.com/future/article/20190207-how-artificial-intelligence-can-help-stop-bullying. Accessed 16 July 2022

31. De Vel O, Anderson A, Corney M, Mohay G (2001) Mining e-mail content for author identification forensics. ACM SIGMOD Rec 30(4):55–64

Pivoting Human Resource Policy Around Emerging Invasive and Non-invasive Neurotechnology

Oliver A. Guidetti and Craig P. Speelman

Abstract This chapter discusses a range of novel yet plausible applications of neurotechnology in a future cyber smart city workforce, along with regulatory mechanisms which may be necessary in mitigating the societal challenges each could pose. If the impact of brain stimulation technologies on the cyber workforce of tomorrow does not take a considered approach, a class divide may open, between analysts who use invasive, non-invasive, and no brain enhancement techniques. Cybernetic products such as Musk (J Med Internet Res J Med Internet Res 21(10):e16194, 2019) Neuralink implant could be used to technologically enhance the attentional capacity of cyber network defence analysts (CNDAs) who will be charged with defending virtual threat environments. For example, the first cybernetically enhanced network defence analysts are most likely to appear in the military rather than the civilian space (Parks in Brain chips and the future of human evolution, 2022). Military CNDAs who have a Neuralink implanted into their cortex during their service period, will eventually retire to the civilian sector. Since any attempt at requisitioning the N1 would require a second brain surgery, the military are unlikely to require retiring service members to give up their brain chips. The development of cybernetically enhanced CNDAs should not, however, require surgery to stay competitive in the already underserviced network defence workforce. However, modern neurotechnological wearables, such as the Artinis (Starstim fNIRS, 2019a), (fNIRS—tDCS—EEG, 2019b) Starstim could enable modern human resource policy to pivot around invasive as well as non-invasive methods. Nuanced navigation of invasive and non-invasive neuro-technologies, such as Neuralink and Starstim, will facilitate a smooth, ethically responsible transition to cybernetic enhancement in a future cyber smart city workforce.

O. A. Guidetti (✉) · C. P. Speelman
School of Arts and Humanities, Edith Cowan University, Joondalup, Australia
e-mail: o.guidetti@ecu.edu.au

C. P. Speelman
e-mail: c.speelman@ecu.edu.au

M. Ahmed and P. Haskell-Dowland (eds.), *Cybersecurity for Smart Cities*,
Advanced Sciences and Technologies for Security Applications,
https://doi.org/10.1007/978-3-031-24946-4_3

1 Pivoting Human Resource Policy and Workplace Neurotechnology

Cyber smart cities develop as new technologies gradually integrate the collective intelligence of core elements of society, including government, security, education, healthcare and workforce management [9]. However, the betterments of technologically enhancing the workforce accompany intricate societal challenges that future cyber smart cities will need to manage [2]. For example, improved portability and practical utility to consumer users in the workforce are projected to spur the global market for neurotechnology to grow from $10 billion in 2022 to over $17 billion by 2026 [43].

This chapter is therefore focused on the responsible introduction of neurotechnology into the workforce driving a cyber smart city. This includes an overview of plausible advances in neurotechnology which could find their way into future cyber smart city workforces. These advances are followed by a brief review of guidelines that cyber smart cities may use to navigate the societal hurdles associated with integrating neurotechnology in the workforce. The paper then closes with an ethical case study of the necessity of these guidelines, derived from an account of an unconsidered application of neurotechnology in a current workplace, specifically, a school [32, 58].

When [55, 56] first coined the term "homo-cyberneticus", it was in reference to the dynamic interplay between technology, culture, and the human mind. Technological enhancement of the human experience has drastically advanced since Singer. The integration of neurotechnology with the human mind is as much a splinter point in our evolution, as is the capacity to explore and edit our own DNA [14]. Modern advances in neurotechnology manifest a clear example of our species' gradual transition from homo sapiens-sapiens to homo cyberneticus [48].

Advances in modern neurotechnology will offer future cyber smart city workforces the capacity to cybernetically enhance their performance on the job, beyond the capability of their non-augmented peers. At first, cybernetically enhanced workers are more likely to come from a military rather than civilian background [45]. For example, network defence analysts could use a product like [39] robotically implanted Neuralink N1 to gain a tactical edge in cyber warfare exercises over red hat adversaries. However, removal of the N1 at the end of a network defence analyst's service period would require a second surgery after the device is implanted, which may not be desirable to both parties. Retired military personnel with an invasive cybernetic implant will therefore eventually return to the civilian sector. This could lead to a gradual accumulation of cybernetically enhanced employees in the workforce with augmentations that distinguish them from their non-enhanced peers. Over time, this could lead to a class divide in a cyber smart city workforce, between those who have undergone cybernetic enhancement, and those who choose otherwise. Therefore, although cybernetically enhancing network defence analysts could provide a human factors advantage above non-enhanced adversaries, remaining competitive in this or any other subset of the workforce should not require weighing up a surgery. However, wearable products, such as Artinis [4, 5] Starstim could provide a cyber smart city workforce with a non-invasive alternative for operators who do not want to undergo surgical enhancement. Though this example is not a current concern, it demonstrates the necessity for forward thinking in developing ethical and responsible human resource policies for cybernetically enhanced employees of the future. This chapter outlines a series of novel but plausible future applications of neurotechnology to a future cyber smart city workforce, followed by a discussion of regulatory mechanisms that could help manage the societal challenges each could pose.

Direct to Consumer cybernetic neurotechnology (DTCCN) specifically refers to devices that do not need an intermediary physician's guidance to use as a means of enhancing cognitive performance [63]. Current DTCCNs have not yet demonstrated a level of effective neuromodulation that would spark widespread public use of the technology. However, by 2024, the consumer neurotechnology market has been projected to grow to $15.1 billion [37]. Therefore, increasing advances in AI neurotechnology may lead to future DTCCNs that do offer individuals a level of cognitive enhancement that sparks widespread use [63].

The roll out of future DTCCNs to consumers in the workplace will force human resource policy to pivot around distributive justice. DTCCNs have the potential to exacerbate socioeconomic inequalities, due to differences in the pricing of cybernetic products [34]. The scope of bodies such as the U.S. Food and Drug Administration (FDA) or the Australian Therapeutic Goods Administration (TGA) may need to expand to provide future cyber smart cities with sufficient regulatory protection against socioeconomic inequality exacerbation by DTCCNs as they become adopted by the workforce [16, 59, 63].

Historically, participants in the workforce exercise a range of legal and illicit strategies to improve neurological performance intended to increase their competitiveness in the modern job market [24, 30]. Psychopharmacological Cognitive Enhancement (PCE) in the workplace can take harmless forms, such as caffeine, however employees have also been observed using controlled substances such as amphetamine, methylphenidate and modafinil to enhance workplace performance [30]. For example, illicit PCE use has been demonstrated in approximately 20% of surgeons and academics, and between 15 and 55% of university students [13, 18, 35, 57]. Initially it was thought that, "boosting brain power" was the primary motivation driving employees' workplace PCE use [49]. However, [17] reported a disambiguation of this motivation, and found that illicit PCE use at work was driven by a perception of high pressure to perform, as well as pressure derived from a fear of operating at a disadvantage relative to peers.

The motivation for workplace PCE use reported by [17] could also drive a similarly widespread adoption of neurotechnological enhancement in a future cyber smart city workforce [24]. Moreover, the range of functions which neurotechnology could serve in the workplace is far more extensive than just enhanced attention, which could broaden the range of employees who use them [3, 7, 11, 13, 18, 22, 28, 27, 33, 35, 36–40, 53, 57, 64]. What follows is a summary of a variety of plausible future cybernetic enhancements that could find use in a cyber smart city workforce.

1.1 Enhanced Alertness

The term neuro-doping was first coined by [11] and refers to the cerebral augmentation of capacities such as alertness and motor control, to provide athletes with an asymmetric advantage relative to their peers. Just as [17] pointed out, the fear of operating at a disadvantage relative to peers is a PCE motivator at work, and this could equally drive neuro-doping in future professional athletes. For example, the Pupillary Unrest Index (PUI) is a method which can be used to measure alertness [47]. Zulkify [64] used the PUI to demonstrate that alertness and motor control could be enhanced by transcranial alternating current stimulation (tACS). Zulkify's result raises a set of concerning questions for a future cyber smart city workforce. For example, future iterations of Zulkify's results could be used by professional athletes to enhance performance in sports where alertness and motor control are key to success, such as automotive racing. A tACS headset built into a driver's helmet could conceivably provide a competitive edge in competition over non-augmented peers. The problems raised by that possibility, however, are whether this counts as cheating, and whether that should inform the way future professional athletes are selected, trained, and managed.

Neuro-doping aside, the ability to augment alertness and motor control could have ethically impactful applications, or use cases, in the workforce of a future cyber smart city beyond professional athletes. For instance, police officers could use the tACS battery Zulkify [64] developed, to bolster alertness and motor control, which could provide added protection during the arrest of violent criminals. Between ten and twenty-five percent of Australian police are assaulted each year during the arrest of violent criminals [12, 36]. Most assaults against police officers result in injuries to the torso, head, face and upper limbs, and are most frequently inflicted by fists during a struggle [36]. However, future iterations of Zulkify's results to enhance officers' capacity to remain alert and physically in control of violent assailants could provide a twofold benefit. Not only could this cybernetic augmentation help protect police during a violent arrest, it could also protect the assailant, as their arrest by an officer with an enhanced level of alertness and physical control could mean that an arrest which would have ended in an assault charge does not escalate to such a point.

1.2 Memory Enhancement

The lateral parietal cortex of the hippocampus (LPCH) is a brain region associated with memory performance [60]. Wang et al. [61] demonstrated that memory performance could be enhanced for a twenty-four-hour period by targeting the LPCH with non-invasive, repetitive transcranial magnetic stimulation (rTMS). It may be the case that advancements beyond Wang et al.'s work could lead to the development of non-invasive wearables that enhance memory performance in future cyber smart city workplaces. Such a device would satisfy each of the motivations [17] reported drive workplace PCE use. Hence future cyber smart city workforces may be driven to use memory enhancing rTMS wearables for much the same reasons as PCEs are used today. That is, future employees may be motivated to augment their memory capacity with rTMS based on perceived performance pressure, as well as a fear of operating at a disadvantage compared to peers who choose to use the technology. Moreover, cybernetic memory enhancement may be particularly likely to arise first in roles associated with heavy PCE motivation and use, including surgeons, academics, and university students [13, 18, 35, 57].

1.3 Deception Detection

Cybernetic enhancements capable of detecting lies and deception may see adoption by the legal profession in a future cyber smart city [7, 38]. Modern polygraphs are today considered unreliable lie detectors, as their results hinge on a range of contextual factors, including question format, and operator bias [52]. For example, polygraphs have demonstrated between 55 and 99% accuracy, an unacceptable level of variability for use in a court of law [52]. Instead of polygraphs, however, future cyber smart city courts of law may instead use cortical lie detectors [7, 50]. For example, modern EEG and FNIR systems have been shown to detect deceptive intent with accuracies of up to 86% and 83% respectively [29, 31]. Moreover, hybrid systems which combine EEG with Functional Near Infrared Spectroscopy (fNIRS), such as the Artinis [4, 5] Starstim, have been shown to achieve a far higher deception detection accuracy of 94% [26]. Cortical lie detectors may therefore see more ubiquitous adoption by the legal profession as the technology matures to a level of accuracy equal to or greater than modern polygraphs [7, 38]. Cortical lie detection methods are less intrusive, easier to accommodate on the stand in a court of law and rely on an artificial intelligence, rather than polygraphs which rely on human interpretation [7]. However before that can happen, the artificial intelligence used to drive cortical deception detection must advance beyond current systems, which are known to be subject to racial and gendered biases [44]. As these biases are developed out of the technology, cortical deception detection could see adoption by the legal profession and courts of law, however this should necessarily hinge on formal demonstration that said systems are not biased.

1.4 Insider Threats, Espionage and Law Enforcement

Insider threats refer to hazardous internal employees motivated to use legitimate work privileges to accomplish illegitimate ends [42]. For example, Henderson's review [23] demonstrated that up to 90% of industries in the United States of America are vulnerable to insider threats. Insider threats are an issue that is anticipated to persistently challenge governments and industry well into the future [21, 54]. However, future cyber smart cities could use EEG to detect malicious deceptive intent associated with insider threats to organisations with over 90% accuracy [1, 22, 28, 27]. The workforce of future cyber smart cities could therefore use cybernetic enhancements to assert trust in employees filling sensitive roles [6, 28]. For example, [28, 27] demonstrated that an insider who provides external, unauthorised agents access to sensitive nuclear facilities could be detected using an EEG. Moreover, [1] demonstrated an EEG insider threat and fitness for duty assessment that was 97% accurate, and also generalised beyond the specific nuclear facility example illustrated by [28, 27]. Future cyber smart city workplaces that are sensitive to insider threats could eventually use EEG as a cybernetic method of security vetting employees in roles sensitive to insider threats.

The possibility of neural code reconstruction from activity recorded by invasive technologies, such as [39] N1 Neuralink could be used in future cyber smart city law enforcement, or by espionage operators. Neural reconstruction of images from cortical activity has demonstrated a sufficient level of maturity to make out text, colours, and shapes [53]. Classically, neural vision reconstruction has required costly fMRIs, however Nemrodov et al. [40] more recently demonstrated this using low cost EEG. Nemrodov's [40] contribution to the literature was not just that they demonstrated neural image decoding by EEG instead of fMRI, their method also advanced the level of detail that could be reconstructed, which included facial features and emotional expressions (Fig. 1). In addition, [3, 33] have recently demonstrated the ability to reconstruct speech and sound from neural data recorded by EEG.

Future advances beyond the work of Nemrodov, Luo et al. and Anumanchipalli et al. could combine the ability to decode sight and sound from neural information to produce a video feed. Moreover, future iterations of [39] Neuralink could be used to record sight and sound information from the cortex that can then be reconstructed into a video feed covertly. That is, neurotechnology wearables placed over the scalp are clearly observable, whereas Neuralink is a surgically implanted brain chip that sits covertly over the cortex but within the skull. Covert neural reconstruction of

Fig. 1 (Top) fMRI reconstruction from [53]. (Bottom) EEG reconstruction from Nemrodov et al. [40]

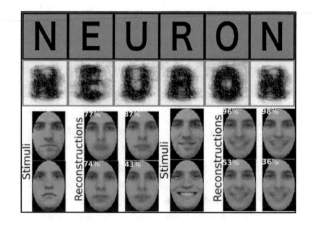

video from sight and sound information recorded with surgically implanted products like [39] Neuralink, could provide immense utility to espionage, or undercover police work. For example, future iterations of the Neuralink implant could be used to record the sights and sounds a foreign agent sees and hears, thus allowing them to record video and audio of restricted facilities, read documents they have not been granted access to, and thus, covertly exfiltrate sensitive protected information. Future cyber smart city workplaces may hence face a day where an X-Ray is required to gain access to secure areas, or to handle sensitive documents. However, this same capability could be used by law enforcement in undercover police work to exfiltrate much the same type of information from criminal organisations in a future cyber smart city.

2 Cybernetic Rights in a Cyber Smart City

Future cyber smart cities will need to pivot human resource policies to accommodate the complex set of challenges posed by integrating neurotechnology into the workplace. For example, the NeuroRights Foundation [41] suggested five "Neuro-Rights", designed to protect society from the misapplication of neurotechnology. These five neuro-rights could inform cybernetic human resource management policy in a future cyber smart city, and include mental privacy, personal identity, free will, fair access to mental augmentation, and algorithmic bias protection.

2.1 Mental Privacy

The right to mental privacy mandates that users' neural data gets stored and processed according to robust cyber security standards [41]. Moreover, the right to mental privacy further stipulates that any sale or commercial transfer of neural data must be rigidly regulated. Finally, the right to mental privacy requires that neural data must be deleted upon the user's request. For example, future cyber smart cities could use EEG to detect insider threats in sensitive workplaces [22, 28, 27, 40]. However, that could lead to a situation in which approval of an employee's security clearance hinges on giving up the right to mental privacy. Similarly, while it is true that the capacity to reconstruct speech from neural code could assist in giving people with disabilities a voice, the technology means their raw thoughts are parsed into data external to the mind [3, 33]. By treating mental privacy as a fundamental principle of system design, future cyber smart cities may find the technology easier to adopt, and thus, benefit from.

2.2   Personal Identity

Boundaries must be drawn between the user's "sense of self" and any cybernetic neurotechnology deployed in the workplace [41]. For example, if neurotechnology connects individuals across a network, there may be a line crossed between that which is the user's consciousness, and that which is derived from an external neurotechnological input.

2.3   Free Will

The right to free will stipulates that the user must retain ultimate control of their decision-making process, unfettered by external cybernetic manipulation [41]. For example, police officers could use a tACS headset to enhance alertness and motor control in order to effect a safer arrest of criminals prone to violence [64]. However, this application could also obscure the free will exercised by officers. The societal benefits of this potential use case of neurotechnology in the workforce will hence rely heavily on the robust, formal validation that users' free will is not obscured during use.

2.4   Fair Access to Mental Augmentation (FAMA)

FAMA is the third right outlined by [41] and mandates guaranteed equality of access to cybernetic neurotechnologies across both national and international levels. By enshrining fair access to mental augmentation technology as a right, FAMA could provide a solution to DTCCN's exacerbation of socioeconomic inequality [34]. For example, the same motivations that drive workplace PCE use could also drive the uptake of rTMS wearables to effect an enhanced memory capacity [17]. Moreover, cybernetic memory enhancement is most likely to arise with professionals where PCEs are already used, including surgeons, academics, and university students [13, 18, 35, 57]. Especially in the case of university students, asymmetric access to cybernetic memory enhancement could lead to drastic divides in the workforce and illustrates why fair access to mental augmentation should be a right for future cyber smart city workforces. However, if access to the technology is fair and accessible across society, this change will be less likely to asymmetrically impact the workforce.

2.5   Algorithmic Bias Protection

Whilst artificial intelligence has been used to effect extensive capability development in neurotechnology, they are prone to the amplification of racial and gendered biases [44, 51]. If these biases are not taken into consideration for each use case, neurotechnology could become a source of tyranny, and lead a future cyber smart city to a 'digital dystopia' [25]. For example, deploying cybernetic deception detection in a court of law should require that the technology's impartiality to race and gender is as robustly and thoroughly demonstrated as any other legally informative scientific implementation. The technology could be of significant utility to the legal profession, so developing solutions to algorithmic biases today will mean that future cyber smart cities of tomorrow derive productive utility from the technology [7, 38].

These five [41] complement a series of thematically similar but more detailed principles outlined by [19] for The Organisation for Economic Co-operation and Development (OECD). Garden et al. detailed nine principles that the OECD should use to inform the integration of cybernetic neurotechnology in the workforce of a future cyber smart city. Garden et al.'s principles are outlined briefly, before being used in conjunction with the five NeuroRights to ethically evaluate a case study of neurotechnology applied to a workplace [32, 58].

2.6   Responsible Cybernetic Innovation

Responsible innovation refers to research, development, and deployment of cybernetic technology which adheres to culturally appropriate ethical standards [19]. This principle will promote prosocial cybernetic management of future cyber smart city workforces by ensuring that the economic incentive of development considers and avoids potential harms and aligns with public values. Similar to [41] these values include, but are not limited to, mental privacy, cognitive liberty, free will and self-determination.

2.7   Authoritative Safety Assessments

Safety Assessments of neurotechnology should be a top regulatory priority. As bodies such as the TGA and FDA expand to include the oversight of cybernetic products, a component of this will include safety standard regulation and formalisation [19]. Development of these standards will necessitate engaging with industry and government stakeholders, researchers and their participants, health care bodies and their patients as well as the broader public. Cybernetic safety standards which prioritise autonomy and which minimise or negate psychophysiological harms will promote public trust in, and wider adoption of, neurotechnology in cyber smart city work-places. Moreover, the TGA, FDA, and similar organisations need to provide both long- and short-term mechanisms of oversight to detect and address unforeseen side effects of cybernetic integration within the workforce. These mechanisms should include a mandate to maintain up-to-date product safety and mental privacy reports [16, 19, 59, 63].

2.8   Inclusivity

Similar to the right to algorithmic bias protection and FAMA outlined by the NeuroRights Foundation [41], the management of neurotechnology in the future cyber smart city workforce needs to promote inclusive policies that enable equal adoption by underrepresented and disadvantaged populations [19, 62]. Presently, neurotechnology research has a tendency for exclusivity. For example, Choy et al. [10] found only five studies out of the 80 they reviewed included black participants in the sample, and of those, none stated if or how their data informed the analysis of results. Phenotypic variability, such as differences in skin tone and hair type, are concerningly endemic biases in electro encephalography (EEG) research [15]. EEG is not the only neurotechnology with inclusivity limitations, for instance, phenotypic biases have also been demonstrated with optical methods of cortical measurement, such as fNIRS [46]. If inclusivity issues in neurotechnology remain relegated to the limitations sections of research articles, this may manifest as wider social issues in future cyber smart cities. That is, if cybernetic enhancement asymmetrically helps one subset of the workforce, to the exclusion of others, future cyber smart cities may see divisive social problems in the adoption of neurotechnology to the workplace. Neurotechnology will do future cyber smart cities little good if the cybernetic capabilities we develop do not consider the implications of limited inclusivity in cybernetic research. However, by addressing the issue of inclusivity today, the workforce of tomorrow may equally benefit from cybernetic enhancement.

2.9   Multidisciplinary Development

Inferring the range of social challenges associated with cybernetic enhancement of the future cyber smart city workforce necessitates an interdisciplinary approach, informed by the brain sciences, humanities, engineering and the legal profession [19]. A multidisciplinary approach will provide a diverse academic capacity to develop and refine best practice standards for the management of invasively and non-invasively enhanced members of the cyber smart city workforce. This diversification of the research space is necessary to inform ethical development of technology and human resource management policies. For example, Zulkify's [64] demonstration of modulated attention and motor performance using tACS could one day be used to give law enforcement an added protection when arresting violent criminals. Ethically and responsibly demonstrating neurotechnology can be deployed in the workplace would require input from a range of bodies aside from the brain sciences. Such a system would need to be engineered in such a way as to be unobtrusive in an officer's day to day work, and be able to recognise its user, so only individual officers could use it. Brain science and engineering aside, police human research ethics committees would also need to provide input into the deployment of such systems. Zulkify's result could thus be used to provide future cyber smart city law enforcement with cybernetic protection during the arrest of violent criminals. Though theoretically useful, reaching that point ethically and responsibly would necessitate a multidisciplinary approach to research, development, and deployment.

2.10 Societal Deliberation

The management of neurotechnology in a future cyber smart city workforce should be based on open dialogue with expert communities as well as the public [19]. Cybernetic workforce management and governance policies should be societally deliberated if the benefits of the technology are to be realised, without incurring harm. Promoting societal deliberation of workplace neurotechnology will improve public literacy and trust, as well as guiding the cybernetic research community to new application spaces.

2.11 Authoritative Oversight

A limitation of the five [41] is the organisation lacks the regulatory authority necessary to react to and manage novel unforeseen issues which may arise in cybernetic workforce integration. Responsible cybernetic integration would most efficiently be overseen by authoritative bodies such as the American FDA and Australian TGA. The FDA and TGA's roles in future cyber smart cities will necessitate their operational expansion to include cybernetic devices [16, 19, 59, 63]. Moreover, oversight of the neurotechnology employed by a cybernetic workforce should be reactive to societal, medical, cultural, ethical, and legal issues which could arise in a future cyber smart city. This reactivity should include horizon scanning, research review, and scenario planning [19].

2.12 Biophysical Data Security

The security of any biophysical data associated with a cybernetic product requires that users are clearly informed about what information is collected, and how it is stored and processed [19]. Moreover, individuals should be able to choose the way their information is used, which includes a mandate to allow access and deletion, that is, the right to be forgotten. Secure biophysical information management will ensure that human resource management policies prevent the misuse of user's information.

2.13 Stewardship

Public trust in cybernetic neurotechnology requires developing a culture of steward-ship across the public and private sectors of a cyber smart city [19]. This stewardship should encourage developing best practice standards for businesses in order to assure users of transparency, accountability, responsiveness, integrity, safety and trustwor-thiness [19]. Moreover, stewardship should include a mandate to identify emergent challenges, issues, or gaps within cybernetic governance systems, and explore solu-tions across regulatory bodies, as well as stakeholders in the public and private sectors.

2.14 Anticipation of Unintended Uses

The potential for unintended use and misuse of neurotechnology should be antic-ipated, by regulatory bodies established specifically to address and regulate such issues. For example, sensitive sights and sounds that a criminal would classify as operationally sensitive could be covertly recorded by implanting an officer with [39] Neuralink, and using it to record audio-visual neural activity that can be reconstructed into video as evidence [3, 33, 40]. However, this same use case could also serve the purposes of insider threats, or espionage. For example, consider the same applica-tion of [39] Neuralink only implanted in a foreign agent instead of a police officer. That agent could plausibly record restricted facilities or exfiltrate documents they are unauthorised to read [3, 33, 40]. Garden et al. [19] advice to the OECD suggested the development of mechanisms external to regulatory bodies, which anticipate plau-sible harmful applications of neurotechnology across the short- and long-term, such as insider threat uses of [39] Neuralink. Pivoting human resource policies around emergent neurotechnology therefore necessitates proactive anticipation of potential unintended uses, to protect future cyber smart cities from harm.

2.15 Ethical Case Study: BrainCo

BrainCo is a Harvard Innovation Lab (HIL) offshoot that builds brain computer interface products, including the Focus 1, or FocusEdu, headset [8, 20]. It is not clear precisely how the Focus 1 works, however the headset purports to combine AI and

EEG to measure levels of concentration. BrainCo conducted an unpublished pilot study of the Focus 1 in 2019, to explore its utility as an educational tool in a sample comprising 50 grade school students, in Xiaoshun central primary School in Jinhua City, [32, 58].

Accounts of BrainCo's 2019 Xiaoshun pilot vary between state and independent media outlets [32, 58]. However the headsets illustrated the importance of guidelines like [19, 41] in ethically deploying neurotechnology in a workplace.

Both state and independent media coverage of the pilot suggested that the headbands were purportedly used to track children's level of concentration in class [32, 58]. The Focus 1 fed the children's concentration data to parents and teachers in 10-min intervals. Parents and teachers then used this information to force children to pay more attention on in-class tasks. Teachers purportedly found that the device increased students' grades and improved the childrens' studious discipline. However, children who scored low on the devices reported being punished by parents [32, 58].

In some regards, BrainCo's Focus 1 pilot illustrate strengths of [19, 41] guidelines. For example, the children did not themselves give consent to wear the devices, as BrainCo obtained consent from parents. The children's mental privacy was therefore not maintained, and the impacts of this were at least significant enough to reinform the way they were treated by their parents. Moreover, because children lost the right to mental privacy, the follow-on effect was a loss of the right to free will of their attention. Perhaps a more ethical method of deploying the Focus1 to a school would have been to use the Focus 1 to help students mindfully regulate, rather than police their levels of attention.

To the Jindong District Education Bureau's (JDEB) credit, they did demonstrate some stewardship and authoritative oversight when they called the pilot study off. For example, the JDEB publicly noted they were motivated to call the study off due to the "heated public opinion" it had caused online, which was effectively a direct response to social deliberation. However, where the JDEB needed to improve as an oversight body was in horizon scanning for issues, such as how the loss of some of the children's mental privacy reoriented the way they were treated by their parents.

In closing, the plausible workplace applications of neurotechnology discussed at the top of this chapter illustrate the mixed bag of benefits and challenges future cyber smart cities will face. Moreover, as BrainCo demonstrated, avoiding the potential for harm necessitates pivoting human resource policies to accommodate the adoption of neurotechnology by the workforce today, at the beginning of the shift towards future cyber smart cities [32, 58]. Cybernetic rights should inform the development of cyber smart city human resource policies needed to navigate the adoption of neurotechnology by the workforce. These should include the right to mental privacy, personal identity, free will, fair access, protection from algorithmic bias, a commitment to responsible, inclusive, and multidisciplinary innovation, authoritative oversight of safety assessments and stewardship of biophysical data, social deliberation, as well as horizon scanning mechanisms to anticipate unintended outcomes.

References

1. Al Hammadi AY, Yeun CY, Damiani E, Yoo PD, Hu J, Yeun HK, Yim M-S (2021) Explainable artificial intelligence to evaluate industrial internal security using EEG signals in IoT framework. Ad Hoc Netw 123:102641
2. Al-Saidi M, Zaidan E (2020) Gulf futuristic cities beyond the headlines: understanding the planned cities megatrend. Energy Rep 6:114–121
3. Anumanchipalli GK, Chartier J, Chang EF (2019) Speech synthesis from neural decoding of spoken sentences. Nature 568(7753):493–498
4. Artinis (2019a) Starstim fNIRS. https://www.artinis.com/starstim-fnirs
5. Artinis (2019b) fNIRS—tDCS—EEG. https://www.artinis.com/starstim-fnirs
6. Balachander S (2021) Assessment of potential security threats from advances in neurotechnology. In: Proliferation of weapons-and dual-use technologies. Springer, pp 77–91
7. Bradshaw R (2021) Deception and detection: the use of technology in assessing witness credibility. London Court of International Arbitration
8. BrainCo (2021) BrainCo Privacy Policy. Retrieved 23/06/2022 from https://brainco.tech/privacy-policy
9. Chen D, Wawrzynski P, Lv Z (2021) Cyber security in smart cities: a review of deep learning-based applications and case studies. Sustain Cities Soc 66:102655
10. Choy T, Baker E, Stavropoulos K (2022) Systemic racism in EEG research: considerations and potential solutions. Affect Sci 3(1):14–20
11. Davis NJ (2013) Neurodoping: brain stimulation as a performance-enhancing measure. Sports Med 43(8):649–653
12. Dawes G, Chong MD, Mitchell D, Henni M (2019) Responding to violent assaults against police
13. DeSantis AD, Webb EM, Noar SM (2008) Illicit use of prescription ADHD medications on a college campus: a multimethodological approach. J Am Coll Health 57(3):315–324
14. Doudna J, Sternberg S (2017) A crack in creation: the new power to control evolution. Random House
15. Etienne A, Laroia T, Weigle H, Afelin A, Kelly SK, Krishnan A, Grover P (2020) Novel electrodes for reliable EEG recordings on coarse and curly hair. In: 2020 42nd annual international conference of the IEEE engineering in medicine & biology society (EMBC)
16. FDA (2019) General wellness: policy for low risk devices. C f D a R Health. https://www.fda.gov: 1–12
17. Forlini C, Racine E (2009) Autonomy and coercion in academic "cognitive enhancement" using methylphenidate: perspectives of key stakeholders. Neuroethics 2(3):163–177
18. Franke AG, Bagusat C, Dietz P, Hoffmann I, Simon P, Ulrich R, Lieb K (2013) Use of illicit and prescription drugs for cognitive or mood enhancement among surgeons. BMC Med 11(1):1–9
19. Garden H, Winickoff DE, Frahm NM, Pfotenhauer S (2019) Responsible innovation in neurotechnology enterprises (2019/05). (OECD Science, Technology and Industry Working Papers, Issue. O Publishing. https://www.oecd-ilibrary.org/content/paper/9685e4fd-en
20. HIL (2016) Venturre Teams Brain Co. Retrieved 23/06/2022 from https://innovationlabs.harvard.edu/current-team/brainco/
21. Hartline Jr, C. L. (2017). Examination of insider threats: a growing concern Utica College]
22. Hashem Y, Takabi H, GhasemiGol M, Dantu R (2015) Towards insider threat detection using psychophysiological signals. In: Proceedings of the 7th ACM CCS international workshop on managing insider security threats
23. Henderson J (2015) The insider threat timeline. https://www.insiderthreatdefense.us/pdfs/ITD%20Insider%20Threat%20Timeline%202-21-14.pdf
24. Hopkins PD, Fiser HL (2017) "This position requires some alteration of your brain": on the moral and legal issues of using neurotechnology to modify employees. J Bus Ethics 144(4):783–797
25. Kehrer K (2021) Humans as data. Press Start 7(1):66–82

26. Khalil MA, Ramirez M, George K (2022) Using EEG and fNIRS signals as polygraph. In: 2022 IEEE 12th annual computing and communication workshop and conference (CCWC)
27. Kim JH, Kim CM, Yim M-S (2020) An investigation of insider threat mitigation based on EEG signal classification. Sensors 20(21):6365
28. Kim CM, Sukb H-J, Yim M-S (2021) Biosignal-based recognition tests to mitigate insider threat in nuclear facilities
29. Kohan MD, Nasrabadi AM, Shamsollahi MB (2020) Interview based connectivity analysis of EEG in order to detect deception. Med Hypotheses 136:109517
30. Leon MR, Harms PD, Gilmer DO (2019) PCE use in the workplace: The open secret of performance enhancement. J Manag Inq 28(1):67–70
31. Li F, Zhu H, Xu J, Gao Q, Guo H, Wu S, Li X, He S (2018) Lie detection using fNIRS monitoring of inhibition-related brain regions discriminates infrequent but not frequent liars. Front Hum Neurosci 12:71
32. Liquan, H., & Yu, D. (2019, 31/10/2019). 金华金东区教育局回应"用头环监测小学生上课走神":暂停用. The Paper. https://www.thepaper.cn/newsDetail_forward_4827137
33. Luo S, Rabbani Q, Crone NE (2022) Brain-Computer interface: applications to speech decoding and synthesis to augment communication. Neurotherapeutics 19(1):263–273
34. Lynch Z (2004) Neurotechnology and society (2010–2060). Ann N Y Acad Sci 1013(1):229–233
35. Maher B (2008) Poll results: look who's doping: in January, Nature launched an informal survey into readers' use of cognition-enhancing drugs. Brendan Maher has waded through the results and found large-scale use and a mix of attitudes towards the drugs. Nature 452(7188):674–676
36. Mayhew C (2001) Occupational health and safety risks faced by police officers. Australian Institute of Criminology Canberra
37. Minielly N, Hrincu V, Illes J (2020) Privacy challenges to the democratization of brain data. Iscience 23(6):101134
38. Mundt JC, Smith JW, Ambroziak G (2022) Ocular-motor deception testing in civilly detained sexually violent persons: an alternative to post-conviction sex offender polygraph testing? Appl Cogn Psychol 36(1):32–42
39. Musk E (2019) An integrated brain-machine interface platform with thousands of channels. J Med Internet Res 21(10):e16194
40. Nemrodov D, Niemeier M, Patel A, Nestor A (2018) The neural dynamics of facial identity processing: insights from EEG-based pattern analysis and image reconstruction. Eneuro 5(1)
41. NeuroRights (2021) The Five NeuroRights. https://plum-conch-dwsc.squarespace.com/mission. Accessed 14 June 2022
42. Oladimeji T, Ayo C, Adewumi S (2019) Review on insider threat detection techniques. J Phys: Conf Ser
43. Paek AY, Brantley JA, Evans BJ, Contreras-Vidal JL (2020) Concerns in the blurred divisions between medical and consumer neurotechnology. IEEE Syst J 15(2):3069–3080
44. Park JE, Lee YK, Hahn S (2021) Racial bias in emotion inference: an experimental study using a word embedding method. In: Proceedings of the annual meeting of the cognitive science society
45. Parks N (2022) Brain chips and the future of human evolution
46. Phan T, Rowland R, Ponticorvo A, Le BC, Wilson RH, Sharif SA, Kennedy GT, Bernal NP, Durkin AJ (2021) Characterizing reduced scattering coefficient of normal human skin across different anatomic locations and Fitzpatrick skin types using spatial frequency domain imaging. J Biomed Opt 26(2):026001
47. Regen F, Dorn H, Danker-Hopfe H (2013) Association between pupillary unrest index and waking electroencephalogram activity in sleep-deprived healthy adults. Sleep Med 14(9):902–912
48. Rekimoto, J. (2019). Homo Cyberneticus: The Era of Human-AI Integration. arXiv preprint arXiv:1911.02637.
49. Sahakian B, Morein-Zamir S (2007) Professor's little helper. Nature 450(7173):1157–1159

50. Satar (2019) Truth about polygraph tests. new straits times, 1–7. https://www.nst.com.my/opi nion/columnists/2019/05/487195/truth-about-polygraph-tests
51. Seyyed-Kalantari L, Zhang H, McDermott M, Chen IY, Ghassemi M (2021) Underdiagnosis bias of artificial intelligence algorithms applied to chest radiographs in under-served patient populations. Nat Med 27(12):2176–2182
52. Shapiro ZE (2015) Truth, deceit, and neuroimaging: can functional magnetic resonance imaging serve as a technology-based method of lie detection. Harv J Law Tech 29:527
53. Shen G, Horikawa T, Majima K, Kamitani Y (2017) Deep image reconstruction from human brain activity. PLoS Comput Biol 15(1):e1006633
54. Simpson CJ (2019) Unauthorized disclosures of sensitive and classified information: a meta-synthesis of leadership support, security policy, and security education, training and awareness within the federal government information security culture. Delaware State University
55. Singer A (1966) Television: window on culture or reflection in the glass? Am Sch 35(2):303–309. https://www.jstor.org/stable/41209373
56. Singer A (1968) Homo-Cyberneticus. Nature 218(5114):901–901
57. Smith ME, Farah MJ (2011) Are prescription stimulants "smart pills"? The epidemiology and cognitive neuroscience of prescription stimulant use by normal healthy individuals. Psychol Bull 137(5):717
58. Standaert M, Yunfan Z (2019) Chinese primary school halts trial of device that monitors pupils' brainwaves. The Guardian 1–4. https://www.theguardian.com/world/2019/nov/01/chinese-pri mary-school-halts-trial-of-device-that-monitors-pupils-brainwaves
59. TGA (2021) How the TGA regulates software-based medical devices. ADO Health. Australian Department of Health: 1–30. https://www.tga.gov.au
60. Wagner AD, Shannon BJ, Kahn I, Buckner RL (2005) Parietal lobe contributions to episodic memory retrieval. Trends Cogn Sci 9(9):445–453
61. Wang JX, Rogers LM, Gross EZ, Ryals AJ, Dokucu ME, Brandstatt KL, Hermiller MS, Voss JL (2014) Targeted enhancement of cortical-hippocampal brain networks and associative memory. Science 345(6200):1054–1057
62. Webb EK, Etter JA, Kwasa JA (2022) Addressing racial and phenotypic bias in human neuroscience methods. Nat Neurosci 25(4):410–414
63. Wexler A, Reiner PB (2019) Oversight of direct-to-consumer neurotechnologies. Science 363(6424):234–235
64. Zulkifly MFM (2020) Modulation of plasticity aftereffects at the sensorimotor cortex by transcranial electrical and magnetic stimulation Georg-August-Universität Göttingen]

Privacy and Ethics in a Smart City: Towards Attaining Digital Sovereignty

Nurul Momen

Abstract Many of today's and tomorrow's smart cities will rely upon services and distributed systems that will apply machine learning and artificial intelligence in their operational logic. These smart applications along with their appliances are data-driven, i.e., their smartness is typically based on the collection, exchange, and the processing of large amounts of user data and user information. For individuals, however, this usage of data may mean be both: (a) an exciting journey to discover the new opportunities of smart cities and (b) an uncertain path to ominous consequences when user data is misused. Hence, the ethical dilemmas and data privacy implications increase when such data is shared across multiple stakeholders, e.g. in distributed artificial intelligence (AI) engineering process, which involves multiple parties. As a result, there is an increasing demand for empowering the users with means, capabilities, and techniques to monitor and eventually to control data disclosure. Users as well as system operators are looking for mechanisms and tools which inform users or operators/stakeholders about the aftermath of the data collection or tools which enforce the data protection and governance requirements in pro-active way and even when conventional privacy enhancing techniques are not applicable. This chapter aims to explore the challenges, requirements, benefits, and functionalities of Digital Sovereignty in a smart city environment. The focus also comprises the necessity and hurdles for engineering, implementation, and development of monitoring and control techniques for future distributed system that can accommodate digital sovereignty when multiple stakeholders are involved and when data is processed in a collaborative way.

N. Momen (✉)
Blekinge Institute of Technology, Karlskrona, Sweden
e-mail: nurul.momen@bth.se

© The Author(s), under exclusive license to Springer Nature Switzerland AG 2023
M. Ahmed and P. Haskell-Dowland (eds.), *Cybersecurity for Smart Cities*,
Advanced Sciences and Technologies for Security Applications,
https://doi.org/10.1007/978-3-031-24946-4_4

1   Introduction

I could not agree more with[1] the fact that the future cities will be 'smart' and living in one of them will introduce numerous and unprecedented conveniences to the inhabitants. However at the same time, I could not help but to worry about the process and the context of injecting 'smartness' into the ecosystem(s) of those people's digital beings. If interacting with a system or an appliance through an interface (e.g., a mobile device), certifies smartness, the resident of such a smart city might face ominous consequences just as frequently as today's data breaches across the globe.

So, how did we manage to get here? How did we acquire such fondness for 'smartness'? Well, I can recall my story in this regard, and I leave it to you—the reader, to decide whether it is relatable or not. I remember my first encounter with a telephone in the early '90s. Back then, phones required a wire. I remember that about a couple of instances when I watched my father anxiously sitting next to a telephone exchange booth after booking a trunk-call. Though hand-held mobile phones started to appear in the movies, it seemed quite a luxury to possess one, at least to myself and to the people living around me. I guess we managed to get past half of the first decade of the new millennium without considering a phone as more than a phone. Of course, there were occasions when new and exciting features were available within a phone: games, torch light, radio, media player and so on. However, it never really managed to occupy our lives as it does today.

Perhaps it would not be an exaggeration to say that nowadays we are in a sense married to our smart devices. We carry them around, pay through them, write on them, keep our secrets in them, talk to them, laugh and cry while talking to them; even we sleep right next to them. Just within a few years of their first appearance, smartphones became very intimate with individuals. At the same time, this tiny device became very powerful compared to what it was a decade ago. Technological advancement and its' exponentially growing miniaturization have truly transformed 'mobile phones' into 'smartphones' by facilitating amalgamation of several components into it. Hence, my personal journey to love 'smartness' commenced, and it progressed in a rapid pace but made subtle changes for doing regular chores.

Over time, we have become (or are becoming) habituated with turning to digital devices to find answers, suggestions, or probable solutions to everyday needs. In order to give it an experimental run, in the recent past (between 2020 and 2021) I tried to carry on living in this modern society with a *non-smart* phone for about 15 months. I have to admit that I struggled a lot to cope up with the everyday needs that are entangled with apps, for instance, digital identification for logging in securely to various accounts. I could not help but to wonder around a couple of things to the least:

---

[1] I began to have a grasp of the overarching thoughts of this chapter during the final days of my Ph.D. journey. At that time, I wrote a small piece as a popular science article to communicate about the significance of my research for the public at large, and I also included it as a "prelude" to the dissertation [14]. Thanks to the editorial team of this book, I was finally able to refine my thoughts, and bring them to a more suitable and relatable context—smart city.

(a) What is it that we ought to sacrifice for the sake of the greater good? (b) Why do we—users need to calculate the trade off between convenience and anomaly?

We try to address the aforementioned queries throughout the remainder of this chapter. First, we highlight on the overarching challenges in Sect. 2. Then the contexts of privacy and ethics are discussed in Sects. 3 and 4, respectively. Afterwards, the regulatory principles are discussed in Sect. 5, which could be used to lay foundation for a framework to address privacy violations and ethical considerations. Finally, we elaborate about a plausible strategy to achieve digital sovereignty in Sect. 6, and then conclude this chapter in Sect. 7.

2 Privacy and Ethics: An Avalanche of Challenges

Let's not start with retelling the history, but I hope you can at least recall how it began with the mobile apps—a freemium[2] approach was adopted within the app ecosystem that required payment for premium service over free ones. Then the nut of behavioral surplus[3] was cracked [3], and the genie of big data began to take over all the other advertising aspects of traditional business entities [6].

Now, our precious time has become the raw material for an economy that is elusive to many [26]. This problem could worsen in a smart city. Why do I sound so ominous here? Let's try to reason with some backgrounds and trends that can highlight such problems.

2.1 Why Is It Challenging?

Nowadays many systems massively rely upon offering an interface through the universal device—mobile devices, and apps are designed and developed to seek attention from us, recurrently. What's the reason behind it? It's simply because of the fact that more screen time of a user translates into revenue for an app.

Can we not just ignore the device? Perhaps the answer is yes, even though it is a 'no' for myself due to having weak control over facing intimidation. Furthermore, the users usually face a cumbersome, and to a certain extent burdensome hurdles which compels them to surrender in favor of the service provider's interest [4, 5, 15]. However, apps do not ignore the user, not even for a narrow window of time interval. They collect data about users' surroundings that contribute to propel the

---

[2] Freemium—Wikipedia: https://en.wikipedia.org/w/index.php?title=Freemium&oldid=964859 444. [Accessed: 2022-07-31].

[3] Behavioral surplus: predictive information about user behavior that is derived from machine intelligence algorithms, which use cumulative data generated from a ubiquitous environment. Hence, human voices, personalities, and emotions can be used in targeted advertising through intervening in the state of play in order to nudge, coax, tune, and herd behavior toward profitable outcomes [26].

data-driven economy. This data collection is so extensive that it is very difficult not to be worried about invasion of my private space, if not impossible. But how can I be so sure about apps' excessive data collection?

I guess, you have already figured it out by now that I happened to be an enthusiast about cyber-physical systems. So, I began to look into the underlying mechanism that allows the app to access user data. In other words, I wanted to find out what happens after the user provides consent by pressing *Allow* or *Accept* button to grant access to personal data.

I often faced a hard time to explain the problem to the audience, and to some extent, even to myself. In the early stage of my research, I found myself ill-equipped with respect to the eloquent skill of communicating science. As a naive opportunist, I opted out for metaphors and other nuances to construct an explanation. Here, I would like to explain the problem with one of them:

> *Let's say, there are apps to wash hands, hypothetically of course. I open the app store and choose an appropriate one for myself. Upon installation, I run it for the very first use and it shows me an interface telling—'Permission required to access Water.' It seems like a legitimate request to me and I press the button—Allow. Consequently, I use the app to wash my hands and afterwards, I put the phone back into my pocket.*
>
> *Wait a second, did I revoke the permission to access water? Is the water still running?*

Though the user leaves the data tap—permission open to hundreds of services in the current context, the means to monitor and to observe privacy-intrusive data collection remain absent [5, 14, 16]. Let me ask you this—what do you usually do when a website asks for your consent to place a cookie on your browser? The cookie-consent scenario for the websites is far more worse than the apps. Websites place hundreds of cookies on our browsers and our online footprints are being tracked and sold to various data brokers. Now, what will happen in a smart city where millions could be living in a pervasive environment? Every device that we ought to interact within such a distributed and connected world, e.g., home appliances, cars, etc., will be responsible for handling user information while delivering convenient services. How are we going to ensure that the information will be gathered and processed ethically while safeguarding individual as well as collective privacy?

I kept wondering about addressing the problem from the other end though: would it not be convenient for the apps and services to just respect user privacy? Why do we, the users, need to calculate the trade-off between convenience and anomaly? Would it matter if the apps and services access more data than needed?

I think, it is a bigger problem than we can anticipate at this moment, because some anomalies can already be noticed that threaten traditional institutions with individual profiling [13]. Currently, at the time of writing this manuscript, the dilemma is even more prominent regarding the contact tracing apps to monitor the spread of a global pandemic [11].

2.2   Is This a Big Deal? If so, Why?

I am not an economist, but certainly there is a concern for the "traditional economy" that we used to have in the good old days. Companies claim private human experience to be a source of free raw materials–that is, behavioral data, which they can and may process by advanced computational techniques to create predictions of our behavior, predictions of what we will do now, soon and later [8]. These derived predictions are then sold to other business entities, which often have regular businesses and they use the new information as a competitive advantage [17]. However, they still buy and sell tangible goods or services—things that you can hold onto, or use/experience/feel. They purchase raw materials, process or produce products as well as services, making sure of the transport and marketing of products, pay taxes and salaries to their employees, so on and so forth. At the end of the day, they can make a few bucks that we call revenues.

Here is an observation to think about: data harvesting companies do not pay for user data, it gets generated as the users carry on with their daily life in this pervasive world [8], which will be even more extensive campaign in the scenario of smart cities. So, their cost for raw material is zero. They do not need to produce their product, i.e., information, it is being derived by running algorithms with a bare-bone minimum cost. They employ a significantly small and highly educated workforce that can leverage the power of a massive capital-intensive infrastructure—compared to traditional business entities with similar equity, e.g., the ratio between Facebook and General Motors is 1:40 [26]. However, these companies are earning money and the amount is just staggering—behavioral surplus had produced a stunning 3590% increase in revenue in less than four years for Google [19].

The world's richest companies have significantly low number of employees. Does it matter? Yes, it does. These companies can and are spending their wealth behind tangible goods and services, which prompts inequalities. Especially now, during the aftermath of a global pandemic, all those small, medium, and even some of the large companies will not be able to survive [25]. Their enterprises cannot carry on while being compelled to go through an indefinite hibernation. In contrary, the big-data companies have this luxury—their low maintenance cost is allowing them to thrive even during a global shut-down state. They will also be able to buy out the little ones—as a 'favor' of course. We will be living in a monopolized economy. The question is: are we ready to accept such inequality within society? or, are we doing so already without even realizing it?

This chapter makes a mere effort to discuss control over the *data-leaking faucets* that can contribute immensely to generate behavioral surplus. It addresses the problem about privacy implication originated from the systems' data access potential. It certainly does not intend to cause impediments for the thriving ecosystem that can bring so much convenience to the future cities, rather it tries to highlight on the proven measures that could handle both individual and collective privacy risks.

Our discussion also includes inspirations from empirical research studies, systems' data collection behavior analyses, visualizing privacy implications, and introducing methods to quantify and to communicate corresponding privacy risks to the user—the inhabitant of tomorrow's smart city. We hope that this effort will contribute to bringing transparency within the ecosystem of systems and thus, encouraging fairness and equality.

3 Privacy

A proper definition for privacy has been hard to come by, at least the one that is being addressed in this complex circumstance which involves numerous entities with fluctuating tolerance thresholds. The Oxford dictionary defines privacy as "a state in which one is not observed or disturbed by other people" [18]. We could also take the "right to be let alone" as the definition of privacy by Warren and Brandeis [23], but the context of this book suits well with the definition provided by Westin back in 1967: *"Privacy is the claim of individuals, groups and institutions to determine for themselves, when, how and to what extent information about them is communicated to others"* [24]. Here, one can correlate the context of a smart city with the significance of the individual right to control visibility of personal information, and associated difficulty to do so successfully, due to variable perceptions and preferences.

How would you address privacy-friendliness within a smart city? Compared to the aforementioned definitions, let's take an inverse route to define privacy-*friendliness*. Let us consider that the violation of privacy is the opposite of privacy-friendliness. However, the violation of privacy is rather debatable because of the fact that it is a subjective concept. Besides self-judgment, there are several other reasons behind diverse concepts of privacy violation; such as geography, culture, and law.

It compels us to lean onto the legal framework and definitions found in literature. In the context of data protection, Solove described violation as an array of harmful activities and placed them into four different categories: (1) information collection, (2) information processing, (3) information dissemination, and (4) invasion [20]. Though 'invasion' is the most common form to express privacy violation, the rest of the forms are equally, if not more important for information privacy protection. Solove argues that collection, processing, and dissemination of information increase the likelihood of potential invasion. For example, the location of a secret army base was exposed in early 2018 from an unlikely source—soldiers used the fitness tracking app *Strava* which was responsible for chronologically *collecting, processing,* and *disseminating* user data that consequently resulted into *invasion.*[4] Hence, considering the other way around, if an event (originated from an intelligent system) is less likely to cause any of the four forms of privacy violation, we can potentially adjudicate that particular activity as a privacy-friendly one. One could think of numerous

---

[4] Fitness tracking app Strava gives away location of secret US army bases: https://www.theguardian.com/world/2018/jan/28/fitness-tracking-app-gives-away-location-of-secret-us-army-bases. [Accessed: 2022-07-31]. More on fitness apps' privacy implications from an empirical study can be found here: [9].

examples and similar scenarios in the context of a smart city, which would resemble the importance of first three phases to protect privacy.

What do you think? Could you think of a few more scenarios?

4 Ethics

It is also rather challenging to define 'ethics' from an unbiased angle, which is probably why philosophers dueled over ethical dilemma throughout centuries. For my personal understanding as a mediocre academic, it is simply the study of moral philosophy that addresses right and wrong behavior. Hence, the role of 'bias' is crucial in this regard, e.g., human rights in different places across the globe. Some might perceive a decision from the government as the righteous one, while it could be criticized with incredulity and indignation in other geographical areas. For instance, social scoring by the Chinese government is typically unacceptable to a European citizen.

Now, why would it matter in the context of modern and digital smart cities? Let's just say that it can appear to be far more complicated! There is a certain possibility that a significant portion of all decisions about our life could be the results from intelligent systems. How are we going to address the bias within it?

We are only beginning to realize some of the dire consequences from such data driven decision support systems. To the date, the European Union is the pioneer to highlight a vision to bring such issues under legal jurisdiction—the Artificial Intelligence (AI) Act in Europe.[5] In short, it will ban activities of a system that could project *unacceptable risks*, e.g., profiling individuals. It also proposes to monitor and regulate activities of intelligent systems that might cause *high risks* to the citizens, e.g., CV scanning tool.

At this point in time, we can only speculate about the depth, variety, and range of ethical matters for the numerous scenarios of a smart city. In the light of current affairs, a couple of scenarios are briefly discussed below.

4.1 Ethical Consideration for the Contact Tracing Apps

COVID-19 pandemic ran havoc throughout the globe. In order to combat it, many nations opted for contact tracing apps for keeping track and limiting the spread of the virus. However, their privacy implications came under scrutiny [11, 22], and various nations had to scramble for preventive measures, e.g., European Union. On the other hand, many nations carried on their agenda anyway, e.g., China, while ignoring such fundamental right of individual privacy.

---

[5] https://artificialintelligenceact.eu/; Accessed: 2022-08-24.

This is a good example of ethical consideration at the highest level of the corresponding governing bodies. One could argue in favor of both ends, while banking on the local, collective, and individual perspectives. Are we ought to sacrifice individual privacy for the sake of public health benefits? Which one would be your ethical stand point, and why? I suppose, such debate will not be an easy one to put together. Anyway, I leave it to you to ponder upon.

4.2 Ethical Dilemma for the Decision Support Systems

Here is a sad, but thought-provoking incident that caused debates across many isles of technology arena. Could you recall what happened to the Germanwings Flight 9525?[6] The co-pilot, who had a medical history of psychological problems, intentionally crashed the plane while killing everyone onboard. According to the investigators, his web searches included phrases like—'ways to commit suicide' and 'cockpit doors and their security provisions'.

Now, we may focus our debate on several aspects, e.g., (a) should the healthcare provider disclose patient information as a precaution in critical cases? (b) should the search engine operate with an alarm system that could potentially prevent such calamity?

These questions come with immensely complex decisions to deal with, specially for the governing bodies who are mandated to answer to the ethical commission. I suppose, such scenarios will be even more abundant in the smart city where automated systems are expected to replace human actors. For example, who are we going to prosecute in case of a road accident involving an autonomous car? Should police arrest the programmer who wrote code for the pedestrian detection mechanism?

5 Taking Refuge Behind the Regulatory Principles

Certainly, we have a long way ahead of us before moving into a smart city. However, we also have some well-established regulatory framework under which one could take refuge, e.g., Europe's General Data Protection Regulation (GDPR) [7]. This could play a vital role in laying foundation for the framework to address and to assess privacy challenges as well as ethical dilemmas. Here, we discuss some key principles from the GDPR, which we believe, could be the starting point for a data-driven ecosystem of a smart city.

GDPR was the pioneer to mandate safeguarding user privacy [7]. Several other countries (e.g., Brazil) followed soon after GDPR came into effect. As we have seen general improvement of privacy and data protection scenarios due to the hefty fine

---

[6] Germanwings Flight 9525; https://en.wikipedia.org/wiki/Germanwings_Flight_9525; Accessed: 2022-08-23.

from this regulatory body, it could be used as the foundation while building robust systems in a smart city. Furthermore, Europe's AI Act is on the horizon, and it can also play a pivotal role in shaping and defining the framework that embodies ethics and privacy.

One would expect that many of the aspects of GDPR are now implemented and this should show in the code. Such expectation should be reflected in the software because of the enforcement of expensive violation sanctions. In theory, the observed data collection and processing behavior of apps should have adapted to the GDPR either (or both) by improved privacy statements and consent collection interfaces or through software updates that changed functionality. However, it is far from being true, which resulted into heavy fines for several entities. In particular, mobile apps have been known to extract large amounts of personal data [2, 10]. This certainly validates the necessity for continuous monitoring and inspection of information systems.

5.1 Inspirations from the Existing Regulatory Framework

As discussed earlier, we bank on the definition from Solove, and put emphasis on the first three stages of privacy violation. In order to do so, we lean onto various requirements on personal data collection and processing that were introduced by the GDPR. As we have observed decent improvement of user privacy in Europe incited by the regulatory shift [16], recommend a few of these requirements that are discussed below.

5.1.1 The Principle of Purpose Specification

It requires any processing of personal data to be bound to a declared purpose. Article 5-1(b) of the GDPR states that personal data shall be *collected for specified, explicit and legitimate purposes and not further processed in a manner that is incompatible with those purposes; further processing for archiving purposes in the public interest, scientific or historical research purposes or statistical purposes shall, in accordance with Article 89(1), not be considered to be incompatible with the initial purposes ("purpose limitation")* [7]. The choice of requested access to user data should, therefore, correlate to the system's functionality and the privacy policy. Moreover, the use of such consent should be bound to its propose—for example, the MICROPHONE permission for your stereo can be used for purpose *A* only, not for purpose *B*.

5.1.2 The Principle of Data Minimization

It requires personal data processing to be reduced to the minimum amount necessary to fulfill the app purpose. Article 5-1(c) of the GDPR states that personal data shall be *adequate, relevant and limited to what is necessary in relation to the purposes for which they are processed ("data minimization")* [7]. Thus, the use of data collection privilege should be restricted to the minimum needed to deliver a transaction with the particular service.

5.1.3 The Principle of Transparency

It requires all personal data processing to be clearly transparent to the data subject. Article 5-1(a) of the GDPR states that personal data shall be *processed lawfully, fairly and in a transparent manner in relation to the data subject ("lawfulness, fairness, and transparency")* [7]. For the systems operating in a smart environment, transparency means that information about the kind of data accessed, the frequency and the amount of data extracted should be available.

5.1.4 Data Protection by Design

This is a principle that requires apps to respect privacy from the start, with safe configurations and minimum necessary data processing. Article 25-2 of the GDPR highlights the requirement: *the controller shall implement appropriate technical and organisational measures for ensuring that, by default, only personal data which are necessary for each specific purpose of the processing are processed. That obligation applies to the amount of personal data collected, the extent of their processing, the period of their storage and their accessibility. In particular, such measures shall ensure that by default personal data are not made accessible without the individuals' intervention to an indefinite number of natural persons* [7]. The implications for collecting user data are that the minimum number of privileges should be requested and a lower number of actual ones should be used in such ways that they reduce the loss of personal data by default.

5.1.5 Data Protection Impact Assessment

This is mandatory for high-risk or high-magnitude data processing or handling of sensitive personal data may uncover risks for data subject privacy. Article 35-7(a–d) of the GDPR elaborates on the criteria: *the assessment shall contain at least: (a) a systematic description of the envisaged processing operations and the purposes of the processing, including, where applicable, the legitimate interest pursued by the controller; (b) an assessment of the necessity and proportionality of the processing operations in relation to the purposes; (c) an assessment of the risks to the rights and freedoms of data subjects referred to in paragraph 1 (Art. 35-1); and (d) the measures envisaged to address the risks, including safeguards, security measures and mechanisms to ensure the protection of personal data and to demonstrate compliance with this Regulation taking into account the rights and legitimate interests of data subjects and other persons concerned* [7]. GDPR compliance activities may uncover such risks materializing as a consequence of excessively broad or deep data collection campaign run by system within a smart city. Consequently, risk reduction will decrease the number of data types consumed and shown.

5.1.6   Freely Given and Unambiguous Data Subject Consent

This is a precondition for lawful data processing. Article 7-2 of the GDPR emphasizes this aspect: *If the data subjects' consent is given in the context of a written declaration which also concerns other matters, the request for consent shall be presented in a manner which is clearly distinguishable from the other matters, in an intelligible and easily accessible form, using clear and plain language. Any part of such a declaration which constitutes an infringement of this Regulation shall not be binding* [7]. Freely given consent would require permission to access data to be bound to a declared purpose, confirmed by the data subject through consent. Therefore, one can expect privacy-friendly systems to change behavior so that user consent will not be confirmed in bulk from the start, but in more selective and interactive ways.

5.1.7   The Right to Withdraw Consent

It is also an important right for implementing individual privacy preferences. Historically, we have seen interfaces that would just show consenting scheme with a binary (accept/decline) choice collection form to the data subject which was addressed in newer versions of several system (e.g., recent changes of mobile operating systems). Now the right to withdraw consent is preserved in several of these versions and therefore they comply with Article 7-3: *The data subject shall have the right to withdraw his or her consent at any time. The withdrawal of consent shall not affect the lawfulness of processing based on consent before its withdrawal. Prior to giving consent, the data subject shall be informed thereof. It shall be as easy to withdraw as to give consent* [7]. However, the spirit of freely given and revocable consent is yet to be fully captured, because there is no privilege usage monitoring mechanism in the user interface. Such unavailability could influence individual decisions on consents given earlier. Thus, privilege usage monitoring efforts can potentially support informed decision-making.

5.2   Is This Enough?

Software changes in reaction to the new regulation are hard to predict, though. While it has recently been argued that the inertia of regulation adaption that is caused by national exceptions and local adaption of EU data protection traditions predating the GDPR [1], we believe that the threat of fines and public exposure actually has created a momentum to improve software, at least when provided by professional software firms or service providers. However, another argument is presented in [21]: the reaction of software vendors to regulation must not always be in support of regulation. As software can be updated very quickly, and standard practices can be established quicker than new regulations can be enforced, there is a risk that software may intentionally use loopholes, camouflage its compliance with regulation, and pursue its own agenda.

Hence, we can reiterate the fact that we need consistent and contentious inspection of privacy compliance, should we ought to address privacy risks as well as ethical issues in a smart city.

6   Towards Digital Sovereignty

So, where does this leave us? What is there to attain through consideration on ethics and privacy? Well, in my opinion, it is hard to point out the benefits of such measures in the society as long as we have them somewhat integrated in the system. For instance, citizens of the western world may remain oblivious about the significance of 'freedom of press' in their society, while people living in the authoritarian regimes are still suffering due to muzzled press. Similarly, the awareness and benefits are invisible in the context of digital sovereignty in our society. If we manage to attain a fair equilibrium of personal data handling that abides by regulatory checks and balances of individual privacy and mandates ethical justifications, we are more likely to look towards a smart city that can offer digital sovereignty to it's citizens.

Now, let us briefly discuss some features of attaining *sovereignty*, however if possible, avoiding the nitty gritty details of it. Sovereignty is mostly used in correlation to a state's universal right to operate and govern itself within its territory without external interference. If we apply the same principle here, we as individuals should have the supreme authority over our presence, footprint, and actions in the digital sphere. If we fail to achieve that, as discussed in Sect. 1, interests in different aspects may appear fragile, or nullified to a certain extent. If we intend to live in a smart city that is able to offer *protection* both in the physical and cyber domains, the measures required to ensure digital sovereignty are non-negotiable.

7   Conclusion

We have heard this statement quite frequently in the recent past: *privacy* is hard, or is it? According to Hoepman, it's a myth, and he made a valiant effort to debunk them in his book [12]!

Also, we have heard blunt statements regarding ethical considerations, for instance—these are subjective matters and they are too volatile to take into account. In my opinion, this is the reason for which we should be open to discuss ethical considerations *prior to designing, during the development, and post deployment* of any digital system. Furthermore, this is certainly not a one-time effort to launch and get it over with, rather it is a continuous and enduring effort to deal with the volatile and evolving nature of threats in the cyberspace.

References

1. Albrecht JP (2016) How the GDPR will change the world. Eur Data Prot L Rev 2:287
2. Barrera D, Kayacik HG, van Oorschot PC, Somayaji A (2010) A methodology for empirical analysis of permission-based security models and its application to android. In: Proceedings of the 17th ACM conference on Computer and communications security. ACM, pp 73–84
3. Bharat K, Lawrence S, Sahami M (2016) Generating user information for use in targeted advertising. US Patent 9,235,849, 12 Jan 2016
4. Bock S, Chowdhury AF, Momen N (2021) Partial consent: a study on user preference for informed consent. In: International Conference on Human-Computer Interaction. Springer, pp 198–216
5. Bock S, Momen N (2020) Nudging the user with privacy indicator: a study on the app selection behavior of the user. In: Proceedings of the 11th Nordic Conference on Human-Computer Interaction: Shaping Experiences, Shaping Society, pp 1–12
6. Cui Y, Shivakumar N, Carobus A, Jindal D, Lawrence S (2005) Content-targeted advertising using collected user behavior data. US Patent App. 10/649,585, 27 Jan 2005
7. EU Regulation (2016) 679 of the European Parliament and of the Council of 27 April 2016 on the protection of natural persons with regard to the processing of personal data and on the free movement of such data, and repealing Directive 95/46/EC (General Data Protection Regulation). Off J Eur Union:L119
8. Goodman A (2019) Age of Surveillance Capitalism: "We thought we were searching Google, but Google was searching us." Democracy Now Interview (Part One) with Shoshana Zuboff. https://www.democracynow.org/2019/3/1/age_of_surveillance_capitalism_we_thought
9. Hatamian M, Momen N, Fritsch L, Rannenberg K (2019) A multilateral privacy impact analysis method for android apps. In: Annual Privacy Forum. Springer, pp 87–106
10. Hatamian M, Serna J, Rannenberg K, Igler B (2017) Fair: fuzzy alarming index rule for privacy analysis in smartphone apps. In: Lopez J, Fischer-Hübner S, Lambrinoudakis C (eds) Trust, privacy and security in digital business. Springer International Publishing, Cham, pp 3–18
11. Hatamian M, Wairimu S, Momen N, Fritsch L (2021) A privacy and security analysis of early-deployed covid-19 contact tracing android apps. Empir Softw Eng 26(3):1–51
12. Hoepman JH (2021) Privacy is hard and seven other myths: achieving privacy through careful design. MIT Press
13. Kosinski M, Stillwell D, Graepel T (2013) Private traits and attributes are predictable from digital records of human behavior. Proc Natl Acad Sci 110(15):5802–5805
14. Momen N (2020) Measuring apps' privacy-friendliness: introducing transparency to apps' data access behavior. PhD thesis, Karlstads universitet
15. Momen N, Bock S, Fritsch L (2020) Accept-maybe-decline: introducing partial consent for the permission-based access control model of android. In: Proceedings of the 25th ACM Symposium on Access Control Models and Technologies. pp 71–80
16. Momen N, Hatamian M, Fritsch L (2019) Did app privacy improve after the GDPR? IEEE Secur & Priv 17(6):10–20
17. Packard NR (2020) Overlooked history in the age of surveillance capitalism. J Media Econ 0(0), 1–7. https://doi.org/10.1080/08997764.2020.1777556
18. Oxford English Dictionary (2018) Online, March 2018: Privacy, noun. Oxford University Press. http://www.oed.com/view/Entry/151596?redirectedFrom=privacy. Accessed 31 May 2018

19. Securities and Exchange Commission, E (2019) Amendment no. 9 to form s-1 registration statement under the securities act of 1933 for google inc. https://www.sec.gov/Archives/edgar/data/1288776/000119312512025336/d260164d10k.htm
20. Solove DJ (2005) A taxonomy of privacy. Univ Pa Law Rev 154:477
21. Wagner RP (2004) On software regulation. South Calif Law Rev 78:457–520
22. Wairimu S, Momen N (2020) Privacy analysis of covid-19 contact tracing apps in the eu. In: Nordic Conference on Secure IT Systems. Springer, pp 213–228
23. Warren SD, Brandeis LD (1890) The right to privacy. Harvard law review, pp 193–220
24. Westin AF (1967) Privacy and freedom. Atheneum, New York
25. Wikipedia (2020) Companies that filed for chapter 11 bankruptcy in 2020. https://en.wikipedia.org/wiki/Category:Companies_that_filed_for_Chapter_11_bankruptcy_in_2020
26. Zuboff S (2019) The age of surveillance capitalism: the fight for a human future at the new frontier of power

Knowledge Organization Systems to Support Cyber-Resilience in Medical Smart Home Environments

Kulsoom S. Bughio and Leslie F. Sikos

Abstract While real-time/near real-time healthcare solutions for the elderly are becoming increasingly popular in medical smart homes, the IT infrastructure and the sensitive information collected using sensor networks can be exposed to cyber-attacks. The management of such systems requires 24/7/365 monitoring and partial automation. However, the wide variety of data formats used by the various vendors' products in medical smart homes cannot be used directly by software agents for data aggregation and data fusion, let alone to support the overall cybersecurity posture. This can be enabled by capturing the semantics of the components of the sensor networks used in these settings. Network anomalies, sensor failure, and sensor data tampering can be detected automatically by performing automated reasoning. In this chapter, we present an overview of knowledge organization systems in a medical smart home environment. We discuss the cyberattack types that occurred in the medical smart home setting and the role of legislation and standards. Finally, we discuss information security countermeasures for cyber-resilience to improve overall the Medical IoT infrastructure.

1 Introduction

The advancement of information and communication technologies (ICT) [1] and wireless sensor networks [2] have led to the design and development of real-time information systems for decision-making, such as smart homes. Such environments are ideal for, among other things, providing a safe, monitored living space where elder people can live without intervention, and their health records can be maintained by healthcare providers, who will assist them timely in case of an emergency [3–5].

K. S. Bughio (✉) · L. F. Sikos
Edith Cowan University, Edith Cowan University Joondalup Campus, Joondalup, WA, Australia
e-mail: k.bughio@ecu.edu.au

L. F. Sikos
e-mail: l.sikos@ecu.edu.au

Remote patient monitoring, telehealth, telemedicine, eHealth applications, and aged care are some examples of medical smart home environments. In these environments, a continuous stream of real-time information about patients can be transferred to remote servers, i.e., cloud server [6] by medical devices equipped with IoT (Internet of Things) sensors and connected through wireless communication technologies such as Bluetooth or Wi-Fi. Medical devices can detect the patient's condition and behavior by monitoring their blood pressure, heart rate, etc. In case of any abnormal reading (which might be caused by not only a system error but also hacking), a doctor or caregiver will be alerted [7].

Based on World Health Organization (WHO) estimates, the section of the population beyond 60 years of age will increase between 2015 and 2050 from 12% to 22% [8]. Similarly, according to the Australian Law Reform Commission, the population of people over 65 will increase to 22.6% by 2054–55 [9]. As the number of elderly expands and considering the COVID-19 crisis, placing older adults in care centres may not be sustainable, while the management of such systems requires 24/7 monitoring and partial automation. The e-PHI "electronic protected health information" is one of the examples, which recognizes all the health information as a covered entity, which is created, maintained, updated or transformed in electronic form. This is the subset of The Health Insurance Portability and Accountability Act of 1996 (HIPAA) [10], where each covered entity guarantees the confidentiality, availability and integrity of all electronic data by detecting threats and protecting information from impermissible uses or disclosures. Some security requirements are needed to protect and safeguard information from threats and attacks [6, 11–14].

In the era of constant cyber threats, smart home environments are connected through a network and can provide services to humans remotely. At the same time, these critical infrastructures face security breaches due to a lack of standard protocols defined by regulatory bodies and device manufacturers. It is very difficult to maintain cyber-resilience in terms of data security, scalability, heterogeneity, interoperability, and integration throughout such networks due to the exposure of data, and the various types of devices having different functionalities and specifications. The rest of this chapter is structured as follows. The related work about the medical smart home environment is discussed in Sect. 2. Section 3 discusses the cyberattack types in medical smart homes while the role of legislation and standards for cyber-resilience in medical smart homes is described in Sect. 4. Knowledge organization systems for cyber-resilience are evaluated in Sect. 5, followed by countermeasures in Sect. 6 with a summary of work in Sect. 7.

2 Related Work

Many researchers in the domain of healthcare work with real-time information systems, where IoT, medical devices, various sensors, cloud services, etc., work collaboratively for the benefit of patients, and facilitates support from their families and medical assistance remotely. In the literature, researchers from various domains discuss medical smart home applications for the elderly. An ambient assistant system called Ontology-based Ambient-Aware LIfeStyle tutoring for A BETter Health (ontoAALISABETH) [15], for example, provides an ambient assisted living framework to monitor the lifestyle of older people, who are not suffering from major chronic diseases or severe disabilities. It integrates an ontology with a rule-based and complex event processing (CEP) engine for supporting timed reasoning, but it cannot handle conflicting events. The framework proposed by Sharma et al. [16] provides information about corona patients to their local associates, family, colleagues etc., for remote monitoring during the COVID-19 pandemic. They proposed an IoT-powered, ontology-based remote access model and bio-wearable sensor system for early detection of COVID-19 by using 1D Biomedical Signals, which contain PPG,[1] ECG,[2] accelerometer, and temperature. This ontology-based monitoring system analyzes the challenges around privacy and security issues and is also simulated by using a cooza simulator to see the efficiency of the proposed model. Other authors proposed a framework for medical diagnosis based on fuzzy Prolog techniques [17]. This framework is implemented and supported by IoT and data analytics and supports real-time monitoring by exploiting the best features of wearable IoT devices. It also correlates information and data from devices by using a context-aware semantic knowledge base.

Based on human motives, goals, and prioritized actions, Guerrero et al.[18] proposed an argument-based approach for tracking and monitoring older people's complex activities. While ALI (Assisted Living System), an argumentation-based multi-agent system [19], was developed for providing support for behaviour change in mild depression and socially isolated individuals in a home environment. It empowers the individuals with the required feedback or recommendations to change their habits and behaviour through an argumentation-based approach. Negin et al. [20] proposed a framework for early diagnosis of cognitive impairments, activity discovery and scene modeling for aged care. Bennasar et al. [3] proposed a system to support older people to live independently in their own homes by increasing their quality of life in terms of care, safety, and security.

3   Common Cyberattack Types in Medical Smart Homes

In any medical smart home where data is transmitted through communication technologies, attackers can attempt to manipulate, steal, interrupt or transmit data. These actions are possible due to vulnerabilities in IoT smart homes, and as a result, cyberattacks occur and reoccur in these environments. The main types of cyberattacks typical of medical smart homes are discussed in the following sections.

---

[1] Photoplethysmogram.
[2] Electrocardiography.

3.1 Cyberattacks on Medical Devices

In a smart home environment, medical devices are connected through the network to transfer the data and information from patients to the server/hospital. Medical devices are facing challenges in terms of cyberattacks in healthcare. Prime examples are medical imaging devices (MIDs), including MRI[3] machines and CT[4] devices. In 2017, the WannaCry ransomware attack was spread over 150 countries and infected 200,000 medical devices, which included 10,000 hospital devices from UK's National Health Service (NHS) and MIDs that became non-operational after encryption. Furthermore, this attack diverted ambulance routes and turned away patients from hospitals [21]. In April 2018, security threats about defibrillators of Abbott ICDs were recorded. The affected medical devices were made up of in-built programmable computer systems, vulnerable to cyberattacks because they were connected to the hospitals via the Internet. The unauthorized user can easily take control of such devices by using equipment that is available commercially and performing various acts through a defibrillator, which, in the worst case, may even result in patient death [22].

3.2 Cyberattacks in Hospitals

In the healthcare industry, where medical devices are frequently targeted by attackers, hospitals and their Internet-connected equipment such as computers also face a big challenge. In February 2016, a $17,000 (40 Bitcoin equivalent at the time) ransom was reported to be paid by the Hollywood Presbyterian Medical Center. The hackers took over the control of the computer systems of the hospital via malware, and as a result, the institution's staff could not communicate through those devices. The hackers gave back access only after the ransom was paid [23]. A month after, the computer network of MedStar, a nonprofit hospital in the USA, was hacked. The hackers asked for a 45 Bitcoin ransom to return all data safely to the 10 hospitals and around 250 outpatient centers. Due to this attack, they shut down many of the computer systems when they received the message from the attacker. While malware attacks like this may be less destructive for other businesses, in healthcare, they destroy or damage at a high level, because patients' e-health records are disturbed due to these attacks, rendering patients unable to communicate with remote hospitals and not being able to gain medical services provided by them [24].

Due to infrequently updated software and standards, in April 2017, many computers and medical devices in the hospitals of the UK were vulnerable and affected by the EternalBlue exploit tool leaked by The Shadow Brokers-a hacker group, among several other exploits. EternalBlue (CVE-2017–0144) is an exploit of the Windows Server Message Block (SMB) protocol which affected various Windows versions, such as Windows Server 2008, Windows Server 2012, Windows 10, Windows Server

---

[3] Magnetic Resonance Imaging.

[4] Computed Tomography.

2016, Windows Vista, etc. Note that a month earlier, Microsoft issued a critical security patch to fix the vulnerability exploited by EternalBlue. However, without the update, many computer systems remained vulnerable [25].

3.3 Cyberattacks in Aged Care Facilities

According to the Australian Cyber Security Centre (ACSC),[5] aged care and hospitals are the main targets for cybercriminals and there is a potential increase shown due to the COVID-19 pandemic-see the Maze ransomware, for example. In the case of Maze, the malicious actors can encrypt the files or lock useful information and demand a ransom. In this regard, the ACSC provides some guidelines such as backup files, using patching software, installing antivirus, checking updates, etc. [26]. In August 2020, Regis Healthcare, a publicly listed aged care provider announced that it was the victim of a cyberattack. A foreign actor had copied some data from Regis hospitals' IT system and released personal information. The healthcare provider used their backups to maintain business continuity so the attack had no more effect on resident care and service delivery, and did not affect day-by-day operations [27].

4   The Role of Legislation and Standards for Cyber-Resilience in Medical Smart Homes

Smart homes constitute a collection of electronic devices, i.e., computers, smartphones, sensors, as well as software, namely, communication network and medical applications, which provide data from the environment. All these can be subject to cyberattacks for various reasons. Firstly, the devices used for collecting medical data are either using Bluetooth or a cloud network, neither of which is secure. Secondly, the people who use electronic medical devices are not all familiar with technology, so they may not follow protocols or standards. Patient data transferred via means of wireless communication are inherently exposed to cyber threats. Consequently, data replacement, data updating, and data stealing are among the main risks in these systems.

The increasing trends of cyberattacks, and in particular, new waves of ransomware, alarm government agencies and regulators to direct policies and standards to protect critical infrastructures and services, such as that healthcare. The Australian Government's TGA[6] Health Safety Regulation provides some standards[7] and is recognized as suitable for manufacturers to be used as a baseline for meeting regulatory require-

---

[5] https://www.cyber.gov.au/acsc/view-all-content/advisories/2020-013-ransomware-targeting-australian-aged-care-and-healthcare-sectors.

[6] Therapeutic Goods Administration.

[7] https://www.tga.gov.au/sites/default/files/medical-device-cyber-security-guidance-industry.pdf.

ments for the cybersecurity of medical devices. Similarly, the US Food and Drug Administration (FDA) also sets guidelines for medical device manufacturers. The premarket and postmarket guidelines are provided as recommendations for the management of medical device cybersecurity risks throughout the product life cycle (Coventry and Branley, 2018). To ensure secure communication between patients, medical devices, and aged care facilities, the manufacturer must follow the standards and protocols defined by regulations and law. The interaction between medical devices, sensors, servers, gateways, and applications can achieve the highest levels of security if they follow the standard protocols. The reliability of connectivity must be ensured by following health industry standards and protocols defined by the government and high authorities.

5 Knowledge Organization Systems for Cyber–Resilience in Medical Smart Home Environments

Knowledge organization systems can play an important role in healthcare using the Internet of Things (IoT). IoT involves several technologies, such as RFID,[8] WSNs,[9] big data, cloud services etc., to capture and recognize the data for smart home applications. There is a need to organize the data coming from various sources in a structured and standard format, so data can be aggregated and/or fused efficiently. Semantic Web technologies are among the examples to represent and organize the data coming from different sources, such as medical IoT sensors. When the associated semantics of the medical IoT devices and their properties are captured correctly, this data could be managed automatically by a software agent to acquire, query, and reuse the existing knowledge for a distributed environment. It will also enable automated alerts when cyberattacks occur.

To support interoperability, Xu et al. presented an IoT-based system for medical emergency services to collect, integrate, and interoperate IoT data in healthcare [28]. For this purpose, a semantic data model was proposed, which can store and interpret heterogeneous IoT data uniformly, a feature particularly useful in real-time applications. A ubiquitous data accessing method (UDA-IoT) was designed to improve accessibility to data sources by acquiring and processing IoT data ubiquitously. This resource-based IoT data accessing method works effectively in distributed heterogeneous environments for timely access to the data in mobile and cloud computing platforms. Similarly, the authors [29] proposed a semantic model based on the IoT platform (SM-IoT), which provides interoperability between data sources and medical devices, along with contract-based security policies for the confidentiality of patient's health data. The Digital Future project is proposed by Pazienza et al. [30],

---

[8] Radio Frequency Identification.

[9] Wireless Sensor Networks.

combining Semantic Web services with SOA-based principles by using an IoE (Internet of Everything) platform. They worked on interoperability at the semantic level by using middleware components.

To address the heterogeneity of different devices by integrating them and identifying and understanding their application programming interfaces (APIs) to collect their data, researchers [31] proposed a mechanism that constructs the ontologies of the various APIs from both known and unknown medical devices and identifies their syntactic and semantic similarity. Based on syntactic and semantic similarity, the final mapping has taken place to identify the nature of API methods for unknown medical devices. After recognizing the APIs of these devices, their data can be gathered through a generic data acquisition API, which is implemented for collecting the data from all devices. The mechanism was evaluated through a use case to identify the API method that provides the data of an unknown activity tracker, compared with the API methods of two activity trackers whose functionality was already known.

6 Information Security Countermeasures in Medical Smart Homes

In the literature, some researchers deal with countermeasures for cyberattacks that occur in IoT networks for medical devices [32]. In [14] Yaqoob et al., discussed the different solutions for security attacks in medical devices such as attestation-based architecture, isolation-based mechanisms, data flow integrity etc. Ahmed et al. [33] developed an ECU-IOHT dataset for analysing attack behaviour and providing countermeasures for the security community. This is a publicly available dataset for the healthcare domain where anomaly detection is performed by nearest neighbour-based algorithms. Although medical smart homes are very attractive to cybercriminals, the medical IoT devices used in them often do not have publicly disclosed technical descriptions, making it challenging for adversaries to attack and security system provides to protect them. In addition, not available configurations make it difficult or infeasible to perform data aggregation and data fusion. The users, including patients, doctors and healthcare providers have very limited interaction with these devices, and they are often unknown from a security perspective.

By building customized smart home settings, all objects might be fully understood and adequately secured, including devices, equipment, and connections. Persons involved in such a setting must be well trained and familiar with technologies and devices Furthermore, while different organizations make roadmaps to secure critical infrastructure from cyberattacks by providing standards and protocols, without having access to technical specifications, the associated semantics cannot be captured in sufficient detail for querying and automated reasoning. Policymakers need to create models for how to implement the standards and protocols in such environments to protect sensitive data from cyber-criminals. There is also a need to develop organizational systems to support the current challenges such as security, privacy, scalability, interoperability, and data heterogeneity, along with business planning to improve cyber-resilience. Better cyber-resilience techniques and solutions can ensure patients and healthcare facilities trust medical IoT infrastructures. Through this, the adaptability and accessibility of medical devices and services can be enhanced and positive outcomes for patients can be increased.

7 Summary

Cyber-resilience is the key element in medical smart home settings. These smart homes are facing challenges due to proprietary solutions, and a lack of organizational systems that can model the technical and security aspects of the system. Semantic models can be seen as one way to arrange and organize the data in healthcare applications. Such semantic modeling of medical IoT devices and their applications provides a machine-readable device, network, and system description, enabling formal knowledge representation and automated reasoning in medical IoT settings.

References

1. Mansell Robin (1994) Information and communication technology policy research in the united kingdom: a perspective. Can J Commun 19(1):1–11
2. Akyildiz IF, Su W, Sankarasubramaniam Y, Cayirci E (2002) Wireless sensor networks: a survey. Comput Netw 38:393–422
3. Bennasar M, Price BA, Stuart A, Gooch D, Mccormick C, Mehta V, Clare L, Blaine A, Stuart A, Gooch D, Mccormick C, Bandara AK (2019) Knowledge-based architecture for recognising activities of older people. 23rd International Conference on Knowledge-Based and Intelligent Information & Engineering Systems, 159:590–599
4. Hammi B, Zeadally S, Khatoun R, Nebhen J (2022) Survey on smart homes: vulnerabilities, risks, and countermeasures. Comput Secur 117
5. Kamel MBM, George LE (2013) Remote patient tracking and monitoring system. Int J Comput Sci Mob Comput 2:88–94
6. Wazid M, Das AK, Rodrigues JJPC, Shetty S, Park Y (2019) IoMT malware detection approaches: analysis and research challenges. IEEE Access 7:182459–182476
7. Mitchell Lauren L, Peterson Colleen M, Rud Shaina R, Jutkowitz Eric, Sarkinen Andrielle, Trost Sierra, Porta Carolyn M, Finlay Jessica M, Gaugler Joseph E (2020) "It's like a cyber-security blanket": the utility of remote activity monitoring in family dementia care. J Appl Gerontol 39(1):86–98
8. World Health Organization (2021) Ageing and health. pp 5–9. Online available at https://www. who.int/news-room/fact-sheets/detail/ageing-and-health
9. Australian government (2017) Australian law reform commission. Who are older australians? pp 1–8
10. Frey Bruce B (1996) Health insurance portability and accountability Act. The SAGE Encyclopedia of Educational Research, Measurement, and Evaluation 12–14:2018
11. Alromaihi S, Elmedany W, Balakrishna C (2018) Cyber security challenges of deploying IoT in smart cities for healthcare applications. In: Proceedings - 2018 IEEE 6th International Conference on Future Internet of Things and Cloud Workshops, W-FiCloud 2018. pp 140–145

12. Nasiri S, Sadoughi F, Tadayon MH, Dehnad A (2019) Security requirements of internet of things-based healthcare system: a survey study. Acta Inform Med 2(4):253–258
13. Sikos LF (2019) Data science in cybersecurity and cyberthreat intelligence
14. Yaqoob Tehreem, Abbas Haider, Atiquzzaman Mohammed (2019) Security vulnerabilities, attacks, countermeasures, and regulations of networked medical devices-a review. IEEE Commun Surv Tutor 21(4):3723–3768
15. Culmone Rosario, Giuliodori Paolo, Quadrini Michela (2015) Human activity recognition using a semantic ontology-based framework. Int J Adv Intell Syst 8(2):159–168
16. Sharma N, Mangla M, Mohanty SN, Gupta D, Tiwari P, Shorfuzzaman M, Rawashdeh M(2021) A smart ontology-based IoT framework for remote patient monitoring. Biomed Signal Process Control 68
17. Martino BD, Esposito A, Liguori S, Ospedale F, Maisto SA, Nacchia S (2018) A fuzzy prolog and ontology driven framework for medical diagnosis using IoT devices. Advances in intelligent systems and computing, 611:875–884
18. Guerrero E, Nieves JC, Lindgren H (2016) An activity-centric argumentation framework for assistive technology aimed at improving health. Argum Comput 7(1):5–33
19. Guerrero E, Lindgren H, Nieves JC (2013) ALI, an ambient assisted living system for supporting behavior change. VIII Workshop on Agents Applied in Health Care (A2HC 2013). Mci, pp 81–92
20. Negin Farhood, Brémond François (2019) An unsupervised framework for online spatiotemporal detection of activities of daily living by hierarchical activity models. Sensors (Switzerland) 19(19):1–34
21. Mahler T, Elovici Y, Shahar Y (2020) A new methodology for information security risk assessment for medical devices and its evaluation
22. Hassija V, Chamola V, Bajpai BC, Naren, Zeadally S (2021) Security issues in implantable medical devices: Fact or fiction? Sustain Cities Soc 66(April 2020):102552
23. Los Angles Times. Hollywood hospital pays $17,000 in bitcoin to hackers; FBI investigating, pp 1–10
24. Farringer Deborah R (2017) Send us the bitcoin or patients will die: addressing the risks of ransomware attacks on hospitals. Seattle Univ Law Rev 40(3):16–23
25. Mahler T, Nissim N, Shalom E, Goldenberg I, Hassman G, Makori A, Kochav I, Elovici Y, Shahar Y (2018) Know your enemy: characteristics of cyber-attacks on medical imaging devices, pp 2–7
26. Australian Cyber Security Centre (2022) 2020-013 Ransomware targeting australian aged care and healthcare sectors, pp 1–6
27. Cheu S (2020) Cyber attack warning: Providers advised to take precautions
28. Boyi Xu, Da Li Xu, Cai Hongming, Xie Cheng, Jingyuan Hu, Fenglin Bu (2014) Ubiquitous data accessing method in iot-based information system for emergency medical services. IEEE Trans Ind Inform 10(2):1578–1586
29. Dridi A, Sassi S, Faiz S (2018) Towards a semantic medical internet of things. In: Proceedings of IEEE/ACS International Conference on Computer Systems and Applications, AICCSA, Oct-2017. pp 1421–1428
30. Pazienza A, Polimeno G, Vitulano F, Maruccia Y (2019) Towards a digital future: an innovative semantic iot integrated platform for industry 4.0, healthcare, and territorial control. In: Conference Proceedings - IEEE International Conference on Systems, Man and Cybernetics, Oct-2019. pp 587–592
31. Mavrogiorgou A, Kiourtis A, Kyriazis D (2020) Identification of IoT medical devices APIs through ontology mapping techniques. EAI/Springer Innovations in Communication and Computing, pp 39–54
32. Hasan MK, Ghazal TM , Saeed RA, Pandey B, Gohel H, Eshmawi AA, Abdel-Khalek S, Alkhassawneh HM (2022) A review on security threats, vulnerabilities, and counter measures of 5G enabled internet-of-medical-things. IET Commun 16(5):421–432
33. Ahmed G, Mehmood D, Shahzad K, Malick RAS (2021) An efficient routing protocol for internet of medical things focusing hot spot node problem. Int J Distrib Sens Netw 17(2)

Cybersafe Capabilities and Utilities for Smart Cities

Kassim Kalinaki, Navod Neranjan Thilakarathne, Hamisi Ramadhan Mubarak, Owais Ahmed Malik, and Musau Abdullatif

Abstract From the beginning of the 21st century, the entire world suffers from two critical problems: the growth of the world population and the improvement of life expectancy of people owing to the development of healthcare facilities. This has ultimately led to urbanization where a lot of people migrated to cities in search of better prospects. To facilitate those migrating into these cities and with the purpose of better provision of city services, smart cities have emerged thereby connecting everything within the city with the aid of a complex set of technologies. The Internet of Things (IoT) is the fundamental building block of smart cities applied in a variety of smart city solutions, offering real-time information exchange and facilitating ubiquitous connectivity. As IoT is a novel technology that is still in its infancy age and requiring continuous internet connectivity, it paves way for never-ending cyber-attacks targeting smart city services and ultimately endangering the lives of city residents. On the other hand, the security of smart city solutions has always been neglected during the development phase which also endangers the entire city's ecosystem resulting into cyber-attacks from multiple attack vectors. In this chapter,

K. Kalinaki (✉) · M. Abdullatif
Department of Computer Science, Islamic University in Uganda (IUIU), P. O Box 2555, Mbale, Uganda
e-mail: kalinaki@iuiu.ac.ug

N. N. Thilakarathne
Department of ICT, Faculty of Technology University of Colombo, Colombo, Sri Lanka
e-mail: navod.neranjan@ict.cmb.ac.lk

H. R. Mubarak
STEM Education, University of Colorado, Boulder, USA
e-mail: ramadhan.hamisi@colorado.edu

O. A. Malik
School of Digital Science, Universiti Brunei Darussalam, Jalan Tungku Link, BE1410 Bandar Seri Begawan, Brunei Darussalam
e-mail: owais.malik@ubd.edu.bn

© The Author(s), under exclusive license to Springer Nature Switzerland AG 2023
M. Ahmed and P. Haskell-Dowland (eds.), *Cybersecurity for Smart Cities*,
Advanced Sciences and Technologies for Security Applications,
https://doi.org/10.1007/978-3-031-24946-4_6

we are providing a brief review of security enhancement capabilities and utilities that can cope with smart cities for the purpose of improving their security against cyber-attacks and safeguarding the privacy of city dwellers.

Keywords Smart city · Cybersecurity · Safety · Privacy · Internet of things · Cybersafe capabilities and utilities

1 Introduction

The technological revolution that began in the early 21st century has fueled the growth of many industries and has introduced many technologies to the world, among which the Internet of Things (IoT) is prominent owing to its ubiquitous connectivity, allowing every digital object to be connected and exchange information [1]. For the time being, the world is undergoing an IoT evolution that connects everything and everyone, allowing for the integration of Information and Communication Technology (ICT) and physical infrastructures (e.g., transportation systems, physical systems, power grids, and so on). This promising connectivity is laying the foundation for making cities smarter by connecting everything within them, including the citizens [1]. The beginning of the 21st century has been impacted by many challenges such as climate challenges, civil and global wars, growing population, urbanization, and disparity in resource allocation [2–4]. On the other hand, at the same time, there was an intensified growth of many digital technologies such as the World Wide Web (WWW) and the Internet. Altogether these challenges and technological revolution have intensified globalization, as to overcome most of the challenges and their adverse consequences leading to an integrated, intelligent, smarter, and sustainable world to make this world a better place [5].

According to the studies [2–6], it is estimated that around 70% of the world population would live in cities by the year 2050 whereas only 13% of the world population lived in cities in 1900. With this rapid urbanization, the world economies have undergone immense pressure to provide necessities that are needed for the survival of citizens in those cities. Energy consumption, public safety, education, transportation, and healthcare facilities were the key resources that have been highly challenged, owing to this rapid urbanization [2]. This continuous pressurization has led to the need for utilizing technology-driven management of cities which paved the way for smart cities [1–5]. In simple terms, a smart city refers to a community that is focused on sustainability, efficiency, and broad participation in decision-making and service provision, which utilizes intensified communication technologies along with IoT as the main backbone. Nevertheless, smart cities have been established in response to the convergence of digital technology and the significant phenomena of community growth and economic innovation that are needed to sustain in the long run.

The IoT being the backbone of the smart city, helps to boost the growth of smart cities by allowing key stakeholders to connect more and more devices, thereby offering seamless ubiquitous connectivity. The fast growth of IoT services in recent years has driven an ever-increasing rivalry in launching new and creative solutions for smart city applications. In doing so, system developers are often pushed to meet rigorous deadlines to maintain their competitive edge [1–4]. Security and privacy needs are frequently seen as afterthoughts in this rushed development process, to be added to the system afterward as features. As a result, the process produces immature solutions that fail to meet the security and privacy criteria of their intended applications, putting the entire IoT ecosystem and the smart city ecosystem in danger resulting in chaos.

On the other hand, the security and privacy of smart cities have not been treated as an integral and important aspect of smart cities until the large-scale ransomware and distributed denial of service (DDOS) attacks encountered recently, resulting in major worldwide chaos [2]. The consequences of these cyber attacks instilled a sense of suspicion in the IoT, prompting some to accuse it of becoming the Internet of Vulnerabilities [2]. Owing to this mere vulnerable nature, the security and privacy of smart cities are becoming a major concern and many people are interested in discovering innovative ways and solutions to overcoming these ramifications [4–6]. Thus, motivated by the fact that discovering these security and privacy-protective mechanisms protecting smart cities, in the following section we outline the key contributions of this book chapter.

- Following the introduction, in the next section we provide a brief overview of the architecture of a smart city, as it is deemed essential to look into the architecture of a typical smart city before moving into the security and privacy aspect.
- A brief overview of IoT in a smart city is provided, as the backbone of a smart city is made out mostly of IoT, whereas IoT applications in a smart city account for most of the vulnerabilities that exist in smart cities.
- A brief outline of cyber security of the smart cities is provided highlighting the security and privacy aspect of smart cities.
- Following discovery of the cyber security aspect of smart cities next we discuss thoroughly the capabilities and utilities available for enhancing the cyber security of smart cities.
- Finally, the future directions for securing smart cities along with the conclusion will be provided.

The remainder of this chapter is organized as follows. Following the introduction, we provide a brief overview of the architecture of a smart city with a special focus on IoT in Sect. 2 as the IoT constitutes the backbone of a smart city. Next in Sect. 3, we discuss more on the cyber security aspect of smart cities while highlighting security and privacy issues. Thereafter in Sect. 4, we thoroughly discuss the available cybersafe capabilities followed by a comprehensive discussion of cybersafe utilities available for protecting smart cities from cyber threats in Sect. 5. In Sect. 6, we summarize the proposed future security and privacy enhancements of the presented cybersafe capabilities and utilities using blockchain technology and finally, we provide a conclusion of the chapter while highlighting its main strength and weakness.

2   Architecture of a Smart City

The architecture of the smart city is a collation of cyber physical systems (CPS) which are made out of a mixture of digital and physical devices. These CPSs are essentially made out of interconnected physical objects such as a variety of sensing and networking devices for intercommunication. These CPSs in smart cities must do three key tasks: data gathering, determining which operations must be performed, and manipulating physical components [1–5]. In light of CPS, they are prevalently used in various industries such as transportation, energy grids, and healthcare for providing smooth and seamless connectivity and performing real-time operations based on real-time data. On the other hand, according to the studies [4–6], a typical smart city can be apportioned into six dimensions as shown in Fig. 1 [5]. These dimensions are: Smart governance, smart economy, smart living, smart people, smart mobility, and smart environment.

On the other hand, the IoT is a vast network of diverse networked items that have a unique identity and can be referenced using IP or MAC addresses [4–6]. The IoT is a subsidiary of the CPS that is made out of the architecture of the smart city which becomes an integral part of CPS. The IoT devices in smart cities include various sensors used for sensing the environment, actuators, intelligent devices, RFID-enabled devices, and smart mobile devices communicating using de-facto communication protocols. As for the time being, the IoT in smart cities is evolving into a technology that allows for the creation of a system made up of cooperating smart autonomous

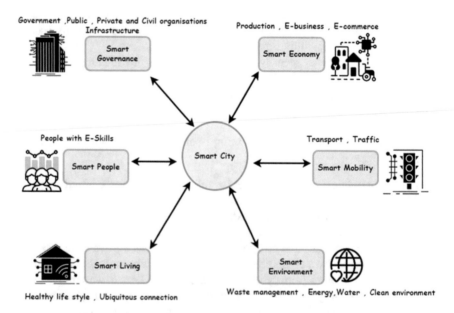

Fig. 1 The six dimensions of a typical smart city

physical-digital devices that are enhanced by sensors and actuators and provide essential processing, storage, and networking capabilities for the smooth operation of smart cities.

In terms of IoT infrastructure used in smart cities apart from sensors and other devices, they consist of communication protocols and APIs (Application Programming Interfaces) used for the collection, aggregation, management, and processing of a large amount of data collated from the city environment which is also known as big data. The implementation might take place on a local or global scale, and it will rely on technologies such as cellular networks, Wi-Fi, and fiber connections for the exchange of data, intercommunication, and connectivity with the Internet. Moreover, cloud computing infrastructures and platforms are also used to deliver flexible cloud-based processing power with big-scale IoT-based CPS.

The underlying communication technologies that provide connectivity to IoT connect the physical city with the data analytics and management units over the Internet. Sensing devices gather data from the city environment, and smart cities modify that data to create a seamless, ubiquitous environment in which data is spread across huge networks and analyzed to produce and give sophisticated intelligent smart services to its people and all stakeholders who are involved in a smart city. According to the studies, it is evident that there is no unique architecture available for smart cities whereas most researchers have referred to the primary IoT architecture as the architecture of the smart city which can be apportioned into three layers; physical layer, network layer, and application layer. For better understanding, the holistic architecture of a typical smart city is presented in Fig. 2 [5].

The physical layer of a typical smart city comprises physical sensing devices which include smart sensing devices, industrial sensors, and wearable devices. These sensors gather data from the physical city and send the gathered data to the processing and, management units in the application layer. These physical sensing devices often belong to the government, private organizations, or individual users. In between the physical and application layer, there is a network layer that is responsible for transmitting gathered data from the physical layer to the application layer with the aid of network infrastructure and underlying communication protocols. The application layer analyzes and processes the obtained sensory data from the physical devices for effective decision-making, using cloud data storage, remote database servers, and specialized control systems. Government institutions, various industries, hospitals, the military, and other approved and authorized bodies have various rights and licenses to examine the underlying information to perform and offer various services.

Further, these institutions will make city-wide rules and regulations based on these inferred data. On the other hand, the smart city also feeds back to alter the actual environment through control and operational components, such as smartphones, based on the decisions made by these processing and management units in the application layer. These control and operational components enhance physical surroundings and improve them to the point where an acceptable quality of life may be achieved in a smart city.

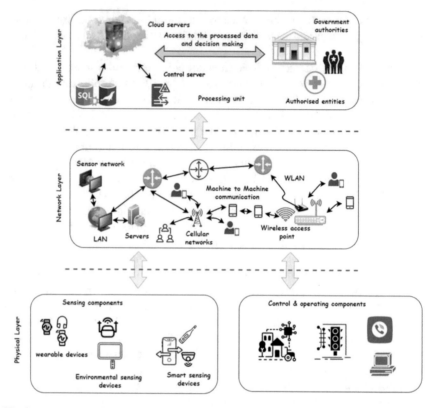

Fig. 2 Architecture of a typical smart city

3 Cyber Security Aspect of a Smart City

Because smart cities are made out of IoT integrated CPS, all the IoT objects in the city are always connected to internal system networks as well as private networks or the internet for real-time exchange of information. This 24/7 connection to the internet always endangers whatever digital devices are connected to the internet in the absence of cyber protection mechanisms. Owing to this vulnerable nature in the absence of cyber security protection mechanism, it would endanger the entire smart city ecosystem where the level of security varies depending on the application and services where the CPS is utilized. Furthermore, the security can also range from physical security to data security while in transmission.

Further, while the smart city applications offer greater flexibility and convenience to city residents, they open up another avenue for malicious cyber-attacks such as system hijacking and Denial of Service (DOS) attacks, jeopardizing every aspect of a smart city. Even though these IoT and associated CPS and eventually the end-user applications offer greater services for city residents, these services often come at a price, increased risk and vulnerability. Thus, the functioning and the operation of a smart city are subjected to the development and deployment of smart city security solutions. To ensure security, even at a minimal level, the following information security requirements have to be met. These also have to be constantly satisfied while configuring devices, services, and key infrastructure in the smart city.

- **Confidentiality**: This relates to the avoidance of information disclosure to unapproved people, organizations, or systems and protects the underlying infrastructure by preventing unauthorized individuals from accessing the generated data.
- **Integrity**: This is the prevention of falsification, and modification of underlying transmitting network data by unauthorized people or devices, and it includes defense against the manipulation of information by injecting messages, replaying messages, and delaying messages on the network.
- **Availability**: This makes sure that only authorized entities may access data, services, and other resources when they are needed.
- **Authenticity**: This security measure is designed to establish the reliability of a transmission, a message, or its author, or to provide a way of confirming a person's consent to access certain data.

4 Cybersafe Capabilities of a Smart City

Comprehensive IoT security solutions which are simple, practical and yet very secure are required to safeguard connected IoT devices in a typical smart city depicted in Fig. 2 above. Instead of proposing a 'super solution' which may fail to work, these solutions are far more effective and different service providers and original equipment manufacturers (OEMs) can easily and widely deploy them. The following section describes the capabilities of such solutions for the security of smart cities.

4.1 Secure Boot and Firmware Integrity

Secure boot deploys techniques based on cryptographic code signing which guarantees that an IoT device only is capable of executing code generated by the device's original equipment manufacturer (OEM) or a trusted party. This technique ensures the prevention of attacks on the configured IoT devices by refusing to execute the program containing the unsigned malware such as worms, viruses, and pre-boot malware. In the end, hackers are restricted from changing the firmware with any other malicious versions of instruction sets [7].

For smart city devices, a secure boot is a required technological capability that is capable of guaranteeing the integrity and authenticity of software packages and also prevents the unsigned code from being executed [8].

4.2 Security Monitoring, Analysis, and Response

This involves the automated ability of a communications network to collect, record, and monitor various data emanating from several endpoints or locations and connectivity traffic. It also involves the analysis of the collected data for purposes of identifying possible violations of security and assessing the severity of any detected threats to the network. Once threats and violations have been detected, response measures should be instituted in line with the general security policies. Such measures can include but are not limited to: temporarily disabling and isolating the compromised devices, quarantining, or complete disconnection and removal of those devices.

This capability is particularly crucial for smart cities which majorly consist of several interlinked IoT devices that constantly communicate and share end-user data. As stated earlier, these devices are vulnerable to a wide range of attacks across all the layers in the smart city architecture.

4.3 Secure, Mutual Authentication

Different components of a smart city can communicate with each other across different layers of the architecture through various network communication protocols. Therefore, establishing secure communication depends on the integrity, confidentiality, and non-repudiation features of network security [9, 10]. The secure and mutual authentication capability for smart cities guarantees that the communicating entities (IoT device and service) can prove their identities to each other before data transmission takes place. This process legitimizes the device and helps prevent malicious attacks from fraudulent devices connected to the network.

4.4 Security Lifecycle Management

This security capability allows original equipment manufacturers (OEMs) to manage IoT device security aspects during the period of their usage such as during a cyber disaster and unauthorized new services for scrapped IoT devices. Secure device shutdown guarantees that devices that have been scrapped will not be re-used to connect to a service without clearance from the authority. Also, to guarantee minimal service outage and disruption of end-user experience during a cyber incident, rapid over the air (OTA) device key(s) replacement can be adopted [11].

4.5 Updating and Patching

Software packages on IoT devices from OEMs must be periodically updated and the inbuilt security features enhanced to ensure their proper functioning as well as safeguard them from new and sophisticated attacks from multiple attack vectors. Furthermore, updating and patching enable the identification of vulnerabilities by enterprises and the provision of the means through which they can be resolved [8].

5 Cybersafe Utilities for Securing Smart Cities

Without the means to ensure the necessary acceptable level of security and privacy, it would be meaningless to call smart cities smart. The holistic architecture of smart cities shown in Fig. 2 of Sect. 2 above involves several interconnected devices supporting different city-wide services across different layers such as physical, network, and application. These interlinked devices, capable of sharing user data, run different applications with unique vulnerabilities that can be exploited [12]. Any single compromised device can result in the rest of the devices across the network being compromised via several methods such as man-in-the-middle attacks, social engineering, denial of service, unauthorized remote recording, botnets, ransomware, data and identity theft, parameter Tampering attacks, Trojan attacks, data spoofing attacks and buffer overflow attacks among others [13, 14]. This therefore poses serious and unique security requirements which in the long run, prevent the widespread adoption and application of the many services offered in smart cities.

In this section, a detailed discussion of the cyber-safe utilities for enhancing the cybersecurity of smart cities is presented.

5.1 Intrusion Detection System (IDS) and Intrusion Prevention Systems (IPS)

An intrusion detection system is a technique in cybersecurity that is capable of detecting intruders and attacks from multiple attack vectors in any communication system such as in IoT. An intrusion prevention system on the other hand is a network security tool that continuously monitors and secures the network from any sort of malicious attacks sent from specific hosts and takes action to prevent them. Both IDSs and IPSs come in either hardware or software implementations and are crucial to be considered in their integration into the IoT environment to mitigate IoT-related security threats that intend to exploit IoT-related security vulnerabilities [7, 12, 13].

For smart cities, in which IoT devices are vastly deployed, IDSs can be deployed in smart transportation services and specifically connected vehicles. Here, the IDS can filter the data exchanged between various vehicles by detecting any anomalies.

In this case, the IDS prevents attacks associated with connected vehicles such as distributed denial of services (DDOS), timing attacks, Sybil and blackhole among others [15]. The IDS can also be deployed in smart health services such as smart hospitals where they are used to detect unauthorized access to private health records of patients through false data injection (FDI) as well as illegal traffic [16]. Finally, intrusion detection systems have also been deployed in smart homes to detect malicious communication from outside the home network, monitor the home network activities of smart home devices and trigger alerts on detected suspicious or malicious behavior [17, 18].

5.2 Honeypots

A honeypot is defined as a cybersecurity technique designed in a safe and controlled manner to lure attackers into a computer system or network [19]. The hackers, upon successfully breaking into a system, think they have access to the real system and yet it's a decoy made with the sole purpose of being broken into. The owners of the honeypot are then able to study the different attack vectors and other weaknesses through which attacks can be made on the real system. In smart home systems, the YAKSHA honeypot is often deployed to collect data for analysis and report information regarding the status of the YAKSHA smart installations system. In so doing, it has shown great success in providing good insights on actual attacks that were launched on a home smart system [20].

Several honeypots such as honeyd, honeydv6, conpot, CryPLH, Supervisory Control and Data Acquisition (SCADA), HoneyNet Project, and SHaPe have been explored for purposes of securing smart grids and industrial control systems (ICS) [21]. All of those honeypots have been used in identifying attacks, gathering intelligence on attack strategies as well as misleading hackers from attacking and causing damage to the smart grid infrastructure. Furthermore, the ZigBee honeypots have been implemented in several profiles for smart health, smart energy, smart agriculture, and smart homes through several standards and specifications intended for short-range wireless technologies [22]. Finally, several honeypots have been applied in water systems with varying levels of interaction (low, medium, high and hybrid) which simulated several services such as transmission control protocol (TCP), EtherNet/Internet Protocol (IP), Hypertext Transfer Protocol (HTTP), and File Transfer Protocol (FTP) [19].

5.3 Demilitarized Zone (DMZ)

A demilitarized zone (DMZ) serves as a perimeter network commonly deployed on an organization's border to protect its internal local area network (LAN) from untrusted traffic. As the network expands with time, it is recommended for any large network to create a perimeter security network such as a DMZ to separate the internal network from the outside untrusted world. Typically, DMZ houses internal resources such as web servers, e-mail servers, domain name servers (DNS) and other systems that have some level of accessibility from the outside world. The resources in DMZ have limited LAN access with an interest to ensure that one can only access them via the public network rather than internal LAN [23]. This, therefore, makes it challenging for hackers to gain direct access to internal systems and sensitive organization data.

In smart cities, a DMZ can be deployed in smart healthcare to separate public resources (such as patient application systems) from internal sensitive information (internal network) and thereby protecting patients' records from being accessed by untrusted traffic [24]. Real-life smart city endeavors such as the Aspern smart city research project in Vienna, Austria have demonstrated the application of demilitarized zones to restrict which services have a higher likelihood of exposure to external entities. For instance, users in such a smart city should have access to the directory containing application programming interfaces (APIs) but be restricted from access to specific APIs via a firewall implementation [25].

5.4 Firewalls

Firewalls (software or hardware) are used to monitor all incoming and outgoing traffic to allow "good data" in, but deny or block "bad data" from entering into a device or network. They act as the first line of defense and gatekeeper for all sorts of traffic flowing in and out of a network [26, 27]. Cloud-based firewalls can be configured in a smart healthcare environment to reduce the impact of cyber threats and safeguard cyber-attacks against smart healthcare devices that carry sensitive data and information [27]. Much as global positioning systems (GPS) and vehicular ad hoc networks (VANET) have long been used for the integrity and overall performance of vehicular networks in big cities, firewalls can also be incorporated to secure the smart transportation system and be free from security breach and jamming of the transportation network [28]. Finally, the firewall can also be deployed in a Smart Home system that uses IoT devices such as smart thermostats, cameras, speakers, toothbrushes, and so on to restrict their access by allowing certain traffic and blocking untrusted access from commanding and controlling IoT devices by unauthorized user [29].

In the table below, we present a summary of cybersafe utilities for securing smart cities (Table 1).

Table 1 Summary of the cybersafe utilities

Utility	Smart city dimensions secured	Articles
Intrusion detection systems (IDSs) and Intrusion Prevention Systems (IPSs)	Smart transportation, smart healthcare and smart homes	[15–18]
Honeypots	Smart homes, smart grids, water systems and smart agriculture	[19–22]
Demilitarized zones	Smart healthcare, an entire smart city	[24, 25]
Firewalls	Smart healthcare, smart transportation, smart homes	[27–29]

6 Future Directions

Owing to their abilities to offer intelligent services such as smart transportation, smart grids, smart healthcare, smart homes, smart agriculture, and smart banking, to mention but a few, the implementation of smart cities is not yet widespread mainly due to numerous security-related concerns which have been partly addressed by the above cyber safe utilities and capabilities. The above services run sophisticated applications that require enhanced security capable of handling the huge amounts of data in the smart city network while at the same time improving the quality of the city dwellers' lives. However, many IoT-related security issues are still unresolved in smart cities and the current technologies and methods are unable to fully address them. In this section, a discussion of the future direction for securing smart cities is provided through the adoption of Blockchain technology which has good security enhancements, especially for IoT.

Defined as a decentralized, transparent, traceable & immutable ledger consisting of transnational records in Peer-to-Peer networks [30], blockchain is considered as a solution capable of enhancing security and privacy in smart cities [31]. In its initial stages, blockchain rose to fame as bitcoin whose solution was for the decentralized transfer of digital payments among different parties [32]. In addition to financial sector improvement, there are several applications where blockchain has potential. Fields like the internet of things (IoT), identity management, accounting and auditing, supply chain, healthcare, telecommunications, energy, and several government public services [33] are some of those in which blockchain is applied.

The table below summarizes the proposed security and privacy enhancements for smart city services using blockchain (Table 2).

Table 2 Summary of the proposed security and privacy enhancements for smart city services using blockchain

Smart city service	Proposed blockchain-enabled security and privacy solution / framework / Protocol / Prototype	Brief description	Articles
Smart e-commerce	Proof of Delivery (PoD) framework	This framework deploys Ethereum smart contracts and blockchain technologies allowing for a secure and transparent logistics control and management of tangible assets either between intermediary transporters or through the sole carrier	[34]
Smart e-commerce	Dual-Deposit escrow protocol	This protocol helps in solving the buyer and Seller's dilemma for selling a digital good in which case the dilemma entails the matter of trust for payment as well as genuine digital goods delivery	[35]
Smart transportation	Blockchain-based Intelligent transportation system (B-ITS) framework	Without technical details for real-world smart city implementation, this proposed framework employs a seven-layered blockchain configuration for securing vehicular networks in smart cities. The layers include the physical, data, network, consensus, incentive, contract, and finally the application layer	[36]
Smart healthcare	MedRec	This is a blockchain-based prototype aimed at providing the means through which e-health records are securely stored for medical research. This prototype is capable of addressing patient privacy together with ensuring improved quality and quantity of medical research data. Modification of medical records in this prototype is prevented through a cryptographic hash application	[37]

(continued)

Table 2 (continued)

Smart city service	Proposed blockchain-enabled security and privacy solution / framework / Protocol / Prototype	Brief description	Articles
Smart grid	Blockchain as a cyber layer, agent/aggregator-based microgrid blockchains, and application-specific blockchains	All those security solutions are capable of providing security and privacy in smart city power grids through the applications of cryptographic securitization together with the consensus mechanism. These ensure data immutability already contained in the blockchain	[38]
Smart home	Homomorphic consortium blockchain model for sensitive data privacy-preserving (HCB-SDPP)	This blockchain-based framework was proposed for the smart home system (SHS) to enhance security and privacy through the application of the Paillier encryption mechanism. Upon analyzing its performance, the framework was determined to be very robust, especially in terms of data availability, data security, and ledger storage security	[39]

7 Conclusions

In this chapter, we have outlined a discussion on capabilities and utilities available for enhancing the cyber security of smart cities along with a summary of key future technologies, from our point of view. As the cyber-attacks targeting internet-facing devices are increasing rapidly, the residents and the relevant stakeholders must act immediately to cover up the vulnerabilities and implement or adapt the cyber security capabilities and utilities towards mitigating unforeseen cyber threats.

The major strength of this chapter lies in the fact that the capabilities and utilities presented can easily and widely be adopted by cyber security specialists, service providers and original equipment manufacturers (OEMs) in their quest to ensure the security and privacy of users in smart cities. One key weakness however lies in the inadequate real-life implementations of some of the summarized future technologies for enhancing the security of smart cities using blockchain. This weakness is attributed to the slow-paced implementation of smart cities worldwide which in turn yields less information on the effectiveness and vulnerabilities of the proposed security enhancements.

In summary, through our discussion, what we have understood is security should be an integral part of a smart city and the developers and architects should put more concerted efforts when designing smart city solutions. We believe this chapter would be an ideal guide for researchers and relevant stakeholders who are keen on this area.

References

1. Jin D, Hannon C, Li Z, Cortes P, Ramaraju S, Burgess P, Buch N, Shahidehpour M (2016) Smart street lighting system: a platform for innovative smart city applications and a new frontier for cyber-security. Electr J 29(10):28–35
2. Habibzadeh H, Nussbaum BH, Anjomshoa F, Kantarci B, Soyata T (2019) A survey on cyber-security, data privacy, and policy issues in cyber-physical system deployments in smart cities. Sustain Cities Soc 50:101660
3. Khan F, Kumar RL, Kadry S, Nam Y, Meqdad MN (2021) Cyber physical systems: a smart city perspective. Int J Electr Comput Eng 11(4):3609
4. ABDOULLAEV A (2011) A smart world: A development model for intelligent cities-[the trinity world of trinity cities]. EIS Encyclopedic Intelligent Systems/SMART GROUP
5. Thilakarathne NN, Madhuka Priyashan W (2022) An overview of security and privacy in smart cities. IoT and IoE Driven Smart Cities, pp 21–44
6. Elhoseny M, Thilakarathne NN, Alghamdi MI, Mahendran RK, Gardezi AA, Weerasinghe H, Welhenge A (2021) Security and privacy issues in medical internet of things: overview, countermeasures, challenges and future directions. Sustain 13(21):11645
7. Singh D, Pati B, Panigrahi CR, Swagatika S (2020) Security issues in iot and their countermeasures in smart city applications. Advanced Computing and Intelligent Engineering. Springer, pp 301–313
8. Sookhak M, Tang H, He Y, Yu FR (2018) Security and privacy of smart cities: a survey, research issues and challenges. IEEE Commun Surv & Tutor 21(2):1718–1743
9. Liu N, Chen J, Zhu L, Zhang J, He Y (2012) A key management scheme for secure communications of advanced metering infrastructure in smart grid. IEEE Trans Ind Electron 60(10):4746–4756
10. Khalil U, Malik OA, Hussain S et al (2022) A blockchain footprint for authentication of iot-enabled smart devices in smart cities: State-of-the-art advancements, challenges and future research directions. IEEE Access 10:76 805–76 823
11. Halder S, Ghosal A, Conti M (2020) Secure over-the-air software updates in connected vehicles: a survey. Comput Netw 178:107343
12. Butt TA, Afzaal M (2019) Security and privacy in smart cities: issues and current solutions. Smart technologies and innovation for a sustainable future. Springer, pp 317–323
13. Alli AA, Kassim K, Mutwalibi N, Hamid H, Ibrahim L (2021) Secure fog-cloud of things: architectures, opportunities and challenges. Secure edge computing, pp 3–20
14. Khalil U, Malik OA, Uddin M, Chen C-L (2022) A comparative analysis on blockchain versus centralized authentication architectures for iot-enabled smart devices in smart cities: a comprehensive review, recent advances, and future research directions. SensS 22(14):5168
15. Aloqaily M, Otoum S, Al Ridhawi I, Jararweh Y (2019) An intrusion detection system for connected vehicles in smart cities. Ad Hoc Netw 90:101842
16. Saba T (2020) Intrusion detection in smart city hospitals using ensemble classifiers. In: 2020 13th International Conference on Developments in eSystems Engineering (DeSE)
17. Kesswani N, Agarwal B (2020) Smartguard: an iot-based intrusion detection system for smart homes. Int J Intell Inf Database Syst 13(1):61–71
18. Alsakran F, Bendiab G, Shiaeles S, Kolokotronis N (2019) Intrusion detection systems for smart home iot devices: experimental comparison study. In: International Symposium on Security in Computing and Communication. Springer, pp 87–98

19. Franco J, Aris A, Canberk B, Uluagac AS (2021) A survey of honeypots and honeynets for internet of things, industrial internet of things, and cyber-physical systems. IEEE Commun Surv & Tutor 23(4):2351–2383
20. Kostopoulos A, Chochliouros IP, Apostolopoulos T, Patsakis C, Tsatsanifos G, Anastasiadis M, Guarino A, Tran B (2020) Realising honeypot-as-a-service for smart home solutions. In: (2020) 5th South-East Europe Design Automation, Computer Engineering, Computer Networks and Social Media Conference (SEEDA-CECNSM). IEEE, pp 1–6
21. Dalamagkas C, Sarigiannidis P, Ioannidis D, Iturbe E, Nikolis O, Ramos F, Rios E, Sarigiannidis A, Tzovaras D (2019) A survey on honeypots, honeynets and their applications on smart grid. In: 2019 IEEE Conference on Network Softwarization (NetSoft). IEEE, pp 93–100
22. Dowling S, Schukat M, Melvin H (2017) A zigbee honeypot to assess iot cyberattack behaviour. In: (2017) 28th Irish signals and systems conference (ISSC). IEEE, pp 1–6
23. Nadig D, Ramamurthy B (2019) Securing large-scale data transfers in campus networks: experiences, issues, and challenges. In: Proceedings of the ACM International Workshop on Security in Software Defined Networks & Network Function Virtualization. pp 29–32
24. Ahmed SM, Rajput A (2020) Threats to patients' privacy in smart healthcare environment. Innovation in Health Informatics. Elsevier, pp 375–393
25. Dhungana D, Engelbrecht G, Parreira JX, Schuster A, Valerio D (2015) Aspern smart ict: data analytics and privacy challenges in a smart city. In: (2015) IEEE 2nd World Forum on Internet of Things (WF-IoT). IEEE, pp 447–452
26. van Oorschot PC (2021) Firewalls and tunnels. Computer security and the internet. Springer, pp 281–308
27. Anwar RW, Abdullah T, Pastore F (2021) Firewall best practices for securing smart healthcare environment: a review. Appl Sci 11(19):9183
28. Jain N, Panda S, Agrawal H (2014) Smart firewall integrated intelligent transportation system for security in ubiquitous computing. Int J Emerg Technol Adv Eng 4(1):684–689
29. Haar C, Buchmann E (2019) Fane: a firewall appliance for the smart home. In: 2019 Federated Conference on Computer Science and Information Systems (FedCSIS). IEEE, pp 449–458
30. Yaqoob I, Salah K, Uddin M, Jayaraman R, Omar M, Imran M (2020) Blockchain for digital twins: recent advances and future research challenges. IEEE Netw 34(5):290–298
31. Biswas K, Muthukkumarasamy V (2016) Securing smart cities using blockchain technology. In: (2016) IEEE 18th international conference on high performance computing and communications; IEEE 14th international conference on smart city; IEEE 2nd international conference on data science and systems (HPCC/SmartCity/DSS). IEEE, pp 1392–1393
32. Nakamoto S (2008) Bitcoin: a peer-to-peer electronic cash system. Decentralized Business Review, p 21260
33. Majeed U, Khan LU, Yaqoob I, Kazmi SA, Salah K, Hong CS (2021) Blockchain for iot-based smart cities: recent advances, requirements, and future challenges. J Netw Comput Appl 181:103007
34. Hasan HR, Salah K (2018) Proof of delivery of digital assets using blockchain and smart contracts. IEEE Access 6:65 439–65 448
35. Asgaonkar A, Krishnamachari B (2019) Solving the buyer and seller's dilemma: a dual-deposit escrow smart contract for provably cheat-proof delivery and payment for a digital good without a trusted mediator. In: 2019 IEEE International Conference on Blockchain and Cryptocurrency (ICBC). IEEE, pp 262–267
36. Yuan Y, Wang F-Y (2016) Towards blockchain-based intelligent transportation systems. In: (2016) IEEE 19th international conference on intelligent transportation systems (ITSC). IEEE, pp 2663–2668
37. Azaria A, Ekblaw A, Vieira T, Lippman A (2016) Medrec: using blockchain for medical data access and permission management. In: (2016) 2nd international conference on open and big data (OBD). IEEE, pp 25–30
38. Musleh AS, Yao G, Muyeen S (2019) Blockchain applications in smart grid–review and frameworks. Ieee Access 7: 86 746–86 757
39. She W, Gu Z-H, Lyu X-K, Liu Q, Tian Z, Liu W (2019) Homomorphic consortium blockchain for smart home system sensitive data privacy preserving. IEEE Access 7:62 058–62 070

Cyber Safe Data Repositories

A. N. M. Bazlur Rashid, Mohiuddin Ahmed, and Abu Barkat Ullah

Abstract Nowadays, data is the center point of source to make any business decisions. With the advancements in digital technologies, business organizations gather a large amount of data using different sources. A large amount of data generation is often termed as Big Data and characterized by many V's indicating their volume, variety, velocity, etc. Organizations use different tools to store and manage these data, and a data repository, such as a data warehouse, data lake, data mart, or data cube, is usually used. Therefore, data repositories are the source of data management and analytics, which are the collection of multiple databases in a structured or unstructured format. While data repositories benefit organizations' large collection of data for business analytics, the question comes to cyber safety and security. Without an appropriate cyber security measure, critical data from the repositories may have unauthorized access—leading to business interrupting financial loss and possible business closer. In this book chapter, the vulnerable issues to the data repositories are discussed, and provided measures for making cyber safe data repositories.

Keywords Data repositories · Cyber safe · Cyber security · Data warehouse · Data lake · Data mart · Data cube · Vulnerabilities · Improving security

A. N. M. B. Rashid (✉) · M. Ahmed
School of Science, Edith Cowan University, Joondalup, WA, Australia
e-mail: a.rashid@ecu.edu.au

M. Ahmed
e-mail: mohiuddin.ahmed@ecu.edu.au

A. B. Ullah
School of Information Technology & Systems, University of Canberra, Canberra, Australia
e-mail: abu.barkatullah@canberra.edu.au

© The Author(s), under exclusive license to Springer Nature Switzerland AG 2023
M. Ahmed and P. Haskell-Dowland (eds.), *Cybersecurity for Smart Cities*,
Advanced Sciences and Technologies for Security Applications,
https://doi.org/10.1007/978-3-031-24946-4_7

1 Introduction

A massive amount of data generation is common to modern technologies. The data fundamentally comes from different sources, including the Internet-of-Things (IoT), financial, economic, health, and cyber domains. The different types of data, such as structured, unstructured, and semi-structured, can be stored in different data repositories, such as Online Analytical Processing (OLAP) or Online Transactional Processing (OLTP), based on the business requirements [1]. However, because the massive generation of data and data has business values, i.e., important patterns can be mined using data analysis, nowadays, data are also required to store as raw data [2–7]. Different variety of data can be stored in different data repositories, such as a data warehouse, data mart, data cube, or data lake. Various application industries, including marketing, retail, and healthcare, have effectively exploited data warehouses. Transactional and analytical operations are two common types used in data processing. To manage daily processes, such as online transactional processing, daily data creation, replication, update, and deletion actions are used (OLTP). Data Mart is a smaller-sized alternative to the data warehouse. Data Marts are smaller and easier to build than data warehouses, which are larger and take longer to build. In contrast to data marts, which only contain a fraction of an organization's data, data warehouses house all of it. A data cube can be compared to a collection of stacked, identical 2-D tables. Data that is too complicated to be displayed as a table of columns and rows are represented using data cubes. Therefore, data cubes may have many more dimensions than just three. Data lakes can be compared to central repositories that are boundless collections of all kinds of data. Thus, data lakes have two characteristics: a variety of data and a schema-on-read approach. Schema-on-read, the opposite of the data warehouse's schema-on-write approach, contends that the schema and data requirements should only be adjusted during data querying. Data lakes are made to manage enormous amounts of unstructured data that arrive quickly, as opposed to a data warehouse that mostly holds structured data [8–12].

However, security lapses at well-known companies worldwide make the news daily. These assaults demonstrate how vulnerable data is and how enterprises of all sizes use weak security measures. The health of a firm as a whole depends on the security of data. The financial information, employment data, and trade secrets all require protection. You might lose money and tarnish your reputation if the security were breached. There are actions you may take to prevent making headlines. Therefore, the organization's data or the data repositories are vulnerable to different types of cyber attacks and weak security measures, such as SQL injection attacks, inadequate access control, exploitable database vulnerabilities, malware, DDoS/DoS attacks, fake data, and data privacy [1, 13]. While these may be the vulnerability issues for the data or repositories, they can be protected by taking security measures, including the use of strong passwords, testing regular security, encrypting data and devices, physical database security, backup data regularly and restoring when required, keeping computer systems upgraded by installing patching and assessing vulnerability and penetration testing [12, 14–19].

The rest of the chapter is organized as follows: Sect. 2 introduces different data repositories. Section 3 discusses the vulnerability issues of the data repositories. Section 4 provides some security measures that can protect the data repositories and improve their security. Finally, Sect. 5 concludes the chapter.

2 Data Repositories

This section introduces the popular data repositories, including data warehouse, data lake, data mart, and data cube.

2.1 Data Warehouse

Large dataset analysis is made possible by the usage of data warehouses, which are business intelligence tools and technology. Data warehouses have been used successfully across a range of application fields, including marketing, retail, and healthcare. Data processing often involves two different types of operations: transactional and analytical. Daily data creation, replicate, update, and delete activities are used to manage daily operations, such as online transactional processing (OLTP). Typically, these data kinds are formatted and kept in a SQL database, such as Oracle Database.[1] Unstructured and semi-structured data are processed and stored in NoSQL databases, like MongoDB,[2] whereas structured data is handled and stored in the context of big data [20, 21]. For analytical purposes, Big Data is also chosen, cleansed, integrated, summed up, and transformed based on the description of the data warehouse schema. Currently, data warehouses are the method of choice for supplying analytical data, and they exclusively store transformed data [11].

Data warehouses are created using the multidimensional model, which specifies the analytical axes, dimensions, and subjects or facts. Dimension tables and fact tables are the two types of tables that make up a data warehouse. The fact tables provide answers to the who, what, when, and where queries. However, the dimension tables receive additional data from the databases based on the fields. Data are extracted (E), transformed (T), and loaded (L) in a data warehouse or ETL procedures. In order to run ad-hoc queries and conveniently extract business intelligence, enterprise data from numerous operational databases is gathered into a single data warehouse storage. To support online transaction processing (OLTP), such as daily business transactions, transactional data is kept in operational databases. On the other hand, data correlations are carried out via online analytical processing (OLAP) processes such as data analysis, examining previous data, and analytical systems. The data warehouse, which was created for analytical purposes, processes all the intricate ad

---

[1] https://www.oracle.com/au/database/.

[2] https://www.mongodb.com/.

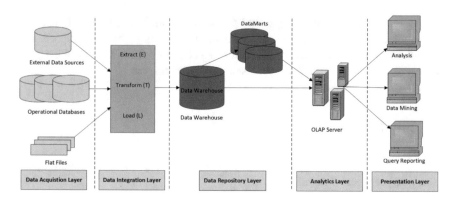

Fig. 1 A conventional data warehouse architecture [9]

hoc queries. The data warehouse can receive batch loads of data. The data that is kept in the warehouse can be subjected to data analytics to help the company make better decisions and gain insightful information [10, 11] (Fig. 1).

Figure 1 illustrates a time-variant data warehouse architecture that is subject-oriented, integrated, and non-volatile for decision-makers. DataMarts, which are compact data warehouses holding a portion of the information gathered from a central data warehouse, are typically found in primary data warehouses. The information pertaining to a certain domain makes up the contents of the DataMarts. Many servers are used to manage and store data in data warehouses. To offer multidimensional perspectives of data, a variety of front-end methods are utilized, including data mining and query reporting [9].

2.2 Data Mart

Data Mart, which is a smaller solution than the data warehouse in terms of size, is one. Data Marts can be created more quickly than data warehouses, which are larger and need more time to develop. Data warehouses host all of an organization's data, whereas Data Marts house only a portion of that data. A department-specific Data Mart can be created by either building a standalone Data Mart or by extracting data from the data warehouse. Subject-oriented, integrated, time-variant and non-volatile data warehouses are all possible. Objects in real time are represented in a subject-oriented data warehouse. To present a single image, an integrated data warehouse combines data from many databases. Data are entered into a time-variant data warehouse based on the time period and saved with the proper timestamps for future analysis and comparison. In a non-volatile data warehouse, as opposed to gathering current transition data, data are collected and placed into the data warehouse at a point in time for analysis. Bitmap indexing and materialized views improve performance in data warehouses. Procedures on data warehouses like aggregate and join operations necessitate a heavy workload of expensive queries. In order to handle complicated ad-hoc queries, notably those involving historical data, which transactional processing databases cannot, data warehouses were largely created [10].

2.3 Data Cube

A data cube can be defined as a three-dimensional (3D) (or higher) range of values during time of time sequence description of an image data. Assessing aggregated data from many angles is a form of data abstraction. Due to the fact that a spectrally-resolved image is represented as a 3-D volume, it is also helpful for imaging spectroscopy. The multidimensional extensions of two-dimensional tables provide another way to define a data cube. It might be thought of as a group of stacked, identical 2-D tables. Complex data are usually challenging to display using a table of columns and rows; data cubes can represent such data. Data cubes may therefore have many more dimensions than merely three. Accordingly, a data cube can be helpful for data with many dimensions for representing various business needs. Each cube's dimension corresponds to a distinct component of the database, such as daily, monthly, or annual sales. The information contained in a data cube enables the analysis of nearly all the data for practically all clients, salespeople, goods, and other variables. A data cube can thus be used to identify patterns and evaluate performance [22].

Data cubes are primarily divided into two groups:

- **Multidimensional Data Cube**: The cube is modeled as a multidimensional array in the framework upon which the majority of OLAP products are constructed. Because this multidimensional OLAP (MOLAP) tools can directly index into the data cube's structure to gather subsets of data, they typically perform better than other alternatives. The cube becomes sparser as the number of dimensions increases. This means there won't be any aggregated data in many cells representing particular attribute combinations. The storage requirements thus rise, sometimes to undesirable levels, making the MOLAP approach inapplicable for exceptionally large data sets with several dimensions. Despite their potential benefits, compression techniques can damage MOLAP's natural indexing.
- **Relational OLAP**: Relational OLAP utilizes the relational database model. The ROLAP data cube is utilized as a collection of relational tables as opposed to a multidimensional array (about twice as many as the number of dimensions). Each table, sometimes called cuboids, represents a certain viewpoint.

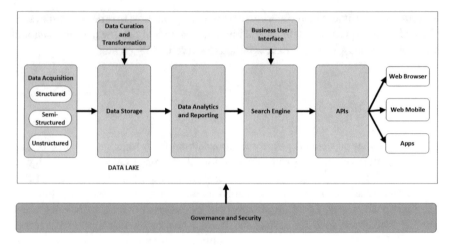

Fig. 2 A conceptual data lake architecture [10]

2.4 Data Lake

Data lakes can be considered central repositories containing all types of data without any schema binding. Consequently, data lakes have two features: (1) A variety of data and (2) A schema-on-read strategy Schema-on-read, which is the reverse of the data warehouse's schema-on-write method, states that the schema and data needs need to be fixed only when data is being queried. Data diversity refers to diverse data types. Data lakes can integrate capabilities for online transaction processing (OLTP), online analytical processing (OLAP), not just SQL (NoSQL), and structured query language (SQL). Unlike a data warehouse that mainly contains structured data, data lakes are designed to handle massive volumes of unstructured data that arrive quickly. As a result, data lakes can use dynamic analytical programs to uncover insights instead of the data warehouse, which uses prebuilt static applications. Similarly, data stored in the lake may be accessed immediately, as opposed to slowly changing data storage in the data warehouse [8]. Real-time analytics, enhanced business intelligence, and very valuable business insights are all provided by data lakes. Figure 2 shows an illustration of a conceptual data lake design [10] (Fig. 2).

3   Are Data Repositories Vulnerable?

Although a data repository has many advantages, there are drawbacks and difficulties. A data repository has many vulnerabilities, particularly in terms of security. Your systems may get slower as the amount of data increases. As a result, your database management system needs to be scalable to handle difficult circumstances. Your entire data may be impacted if a system crashes. To reduce the danger of data loss or intrusion, back up your databases and restrict access to your systems. Additionally, if adequate security measures aren't in place, unauthorized individuals could access your crucial data and abuse it. You must choose the gear and software you employ while dealing with a data repository more carefully. Use only the most effective data warehousing techniques, please. Additionally, involve the important stakeholders in your firm in the process and teach your staff about data security. Hire the top industry experts to look after the repository and ensure your database administration is scalable. You can hire a reputable and trustworthy cyber security service provider for increased security. Use the ETL tools while moving your data repository, please. It would help to create your data marts after creating your data warehouse. A few vulnerability issues that can affect data repositories are described in the following subsections [1, 12–19, 23].

3.1 Database SQL Injection Attacks

SQL injection attacks, which target relational database servers (RDBMS) that employ the SQL language are the most common type of database injection attacks. MongoDB, RavenDB, and Couchbase are examples of NoSQL databases that are resistant to SQL injection attacks but vulnerable to NoSQL injection attacks. Attacks using NoSQL injection are less frequent yet just as harmful. SQL injection and NoSQL injection attacks work by getting through web applications' data entry restrictions to send commands to the database engine, which then exposes the data and structures of the database. A successful injection attack may, under rare circumstances, grant the attacker unrestricted access to the core of a database.

3.2 Accessible Backups

Although a layer of security secures database servers, unprivileged users may have access to the backups of these databases. Unauthorized users risk copying the backups and mounting them on their own servers to extract the sensitive data they contain.

3.3 Inadequate Permissions Management

Database servers are frequently installed in companies with their default security settings in place, and these settings are never altered. This exposes databases to attackers who are aware of the default permissions and are skilled at using them against them. There is also the issue of people abusing lawful permissions by using their access to databases for improper purposes, such as disclosing private data. Since unscrupulous individuals might be aware of the presence of these accounts and use them to gain unauthorized access to databases, the existence of inactive accounts also presents a security concern that is frequently disregarded.

3.4 Hidden Database Servers

Users install database servers at their own discretion to address specific needs due to non-compliance with software installation policies in an organization (or the absence of such standards). As a result, servers show up on the company's network without the security administrators' knowledge. These servers expose the firm to confidential data or reveal security holes that attackers can use.

3.5 Exploitable Database Vulnerabilities

Corporate IT organizations frequently neglect to update their DBMS core software. Therefore, it may take months for businesses to patch their systems even if a vulnerability is found and the vendor publishes a patch to fix it. As a result, vulnerabilities are left open for a long time, which hackers might take advantage of. The primary reasons why DBMSs are not patched include challenges in locating a maintenance window, complicated and time-consuming patch testing procedures, ambiguity on who is in charge of maintaining the DBMS, and an exorbitant workload for system administrators, among others.

3.6 Human Error

Many human activities, such as using weak passwords, unintentional data corruption or erasure, and sharing passwords without appropriate security measures, are the leading reasons for data breaches, for about half of the reportable breaches.

3.7 Buffer Overflow Attacks

A buffer overflow can occur when a system process attempts to store data in large size in a memory block of fixed length. The extra data is usually stored in nearby memory addresses. However, hackers can attack a system using these additional data as base attacks.

3.8   Malware

Software that is designed to exploit flaws or damage a database is known as malware. Any endpoint connected to the network of the database could transmit malware. Any endpoint needs malware security, but database servers require it more than other endpoints due to their high value and delicate nature.

3.9   DDoS/DoS Attacks

A denial of service (DoS) attack is a type of attack when cybercriminal sends a massive amount of fictional requests to a target device or service. An example is the database server for overwhelming it. The server will then frequently crash or become unstable because of the inability to process actual user requests. Fake traffic can be generated when many computing devices are connected to a botnet, which a hacker might already attack—resulting in a distributed denial of service (DDoS) assault. Eventually, this attack can create extremely high traffic and can be challenged to manage without a highly defensive system architecture. Therefore, DDoS prevention services can be scaled up rapidly in the cloud for handling this massive number of DDoS attacks.

3.10   Fake Data

The generation of fake data can create a great challenge to the business because it requires time to identify the root problem and also requires an urgent solution. However, it is difficult for a business to examine each data point. In contrast, using fake data will create erroneous information on a large scale. In addition, fraud data can have false signs, triggering pointless actions and reducing business outputs and operations. The business should be skeptical of data to protect the business operations from such fake data. One of the best strategies to take as an action plan against fake data is using machine learning models to build and tests models using a number of test datasets and regularly validating data sources. This can ensure that business operations run smoothly by observing anomalous activities resulting from fake business data.

3.11   Data Poisoning

Today, there are several machine learning options exist, like chatbots, which have access to vast amounts of data. Such systems can evolve over time when users engage with them. Data Poisoning, a technique for assaulting the training data for machine learning models, is the result of this. It might be considered an integrity assault since the modified training data may affect how well the model predicts the future. Logic corruption, data manipulation, and injection are catastrophic results. Outlier detection is the best method of combating evasion by ensuring a standard data distribution and building an accurate learning model.

3.12 Data Privacy

Data privacy is a significant issue in this digital age. It makes an effort to guard against breaches, hackers, and intentional or unintentional data loss. With cloud-based access management services, organizations must abide by stricter data privacy regulations, including highly rigorous privacy compliance, to increase data protection. The best course of action is to follow some guidelines in addition to installing one or more Data Security solutions. The broad recommendations are to be aware of your data, exercise greater control over your data backups and storage, secure your network from unauthorized access, do frequent risk assessments, and regularly educate your users on data privacy and security.

4 Improving Data Security

A number of security strategies are usually deployed to safeguard a database management system to keep it protected from illegal use and cyberattacks. The database security system can manage the entire system, including its data, potentially damaging, infiltration, and exploitation. Database security consists of many security measures that are built from tools, procedures, and different approaches. A number of security measures can be taken to protect the data repositories are discussed in the following subsections [1, 12–19, 23].

4.1 Testing Security

Protecting database management systems against malicious cyberattacks and unauthorized use involves employing various strategies. Database security solutions are designed to protect against attacks, such as exploitation, intrusion, and damage, but also for the data stored within the database and entire data management system and any application that uses it. You should stroll across your workplace and inspect the desks of your coworkers. If you look closely enough, you can find a sticky note with a password scrawled.

4.2 Strong Passwords

Numerous companies continue to have lax password policies, which leads to crucial accounts with access to delicate and precious data having simple, hackable passwords. Using strong passwords is the first thing you can do to increase your security in this area. Update your passwords at least once every three months, and use reasonably complicated passwords. Never use passwords like "12345" or "admin123". Never save your passwords to a piece of paper or leave them on your computer where someone else could find them.

4.3 Protect the Data Itself

With approximately 90% of security resources going toward firewall technology, many firms focus on protecting the walls around their data. However, firewalls have countless potential workarounds, including through clients, partners, and staff. These people are all capable of getting around external cyber security and abusing critical information. You must be careful to focus your security efforts on the data rather than just the perimeter.

4.4 Updating Programs Regularly

Ensure that your machine has the latest updates and patches. It is frequently better to do this to ensure it is sufficiently secured. The most current update to your security software determines how effective they are. These tools must be updated periodically since ransomware strains and hackers continuously evolve to take advantage of bugs in earlier versions of software.

4.5 Encrypting All Devices, Files, and Backups

In today's society, more and more people prefer to work on their personal or portable electronics. How can you be sure that these instruments are trustworthy? Before saving any data, ensure it is encrypted, and keep it secured during migrations. There is always a possibility that a hacker will gain access to your system, regardless of how strong your protection is. However, there are other security issues with your database besides attackers. Your employees could pose a risk to your business. A malicious or careless insider could potentially view a file they are not supposed to see at any time. When your data is encrypted, it is unreadable by both attackers and employees. A final line of security against uninvited intrusions is provided because they cannot access it without the encryption key. All-important software files, data files, and backups should be encrypted to prevent unauthorized access to your critical information.

4.6   Accounts and Privileges

Take the Oracle database server as an example. Thanks to the Oracle database configuration assistant, the bulk of the default database user accounts are automatically locked and expire after the database installation (DBCA). This won't occur, and the default privileged accounts won't be locked or expire if you install an Oracle database manually. By default, their username and password remain the same. An attacker will initially try to utilize these credentials to log in to the database. A strong, unique password must be set for each privileged account on a database server. An account should be locked and expired if it is no longer needed. Access must be restricted to the barest minimum for the remaining accounts. Only the tables and operations (such as SELECT or INSERT) required by the user should be accessible to each account. Avoid giving users access to all database tables through user accounts.

4.7   Creating Company-Wide Security

Everyone with a username and password is in charge of maintaining data security. IT administrators must routinely remind managers and staff not to divulge login information to any outside party. Everyone has a role in data security; it is not simply the responsibility of the IT team.

4.8   Physical Database Security

It is a best security practice not to share a server between database applications and web apps when the database may contain sensitive data. Even if hosting your website and database on the same server may be easier and less expensive, you are giving a third party control over the security of your data. If you do choose to entrust your database management to a web hosting service, you should ensure that it is a business with a solid reputation for security. Free hosting services should ideally be avoided owing to potential security concerns. Be mindful that an on-premises data center where you keep your database is also susceptible to insider threats and external attacks. Ensure your facility has physical security features, such as locks, cameras, and security guards. Physical server access must be monitored and restricted to those with authorization. Additionally, avoid storing database backups in openly accessible places like temporary partitions, web files, or unprotected cloud storage buckets.

4.9   Insider Threats

Since external dangers are frequently portrayed as the largest and most expensive threats on the news and television. However, it can be simple to picture them. The truth is that the people closest to you have the most capacity to hurt you. Insider threats can be challenging to recognize and control because of their character. It might be as easy as a worker opening an email attachment, which they may think from a reliable source. However, this can trigger a ransomware worm. Threats of this nature are the most frequent and expensive worldwide.

4.10   Time and Money for Cyber Security

The lack of data security continues to be the most significant danger to your IT infrastructure. Many CIOs have acknowledged that you need to invest more time and money in it. All business operations must incorporate cybersecurity, so many large firms with sensitive corporate data frequently hire chief security officers for board-level positions.

4.11   Deleting Redundant Data

Many companies' operations depend heavily on sensitive information, especially healthcare, banking, the public sector, and education. Putting in place information disposal policies helps prevent lost or stolen data in the future. It will be much easier to control your staff from storing redundant data if you have a procedure for destroying, deleting, or otherwise altering it to make it unreadable.

4.12   Backup Data Regularly

To protect from insider and outsider threats, backing up data regularly is now an integral part of IT security policies. A secure and regular data backup can restore service and business operations at any point of failure, accidental file deletion, and even in the case of a ransomware attack that restricts access to the data. However, it should also be ensured that the data backup is stored in secure storage at a greater distance from the main business location.

4.13   Disabling Public Network Access

It is usually common that organizations' data and applications are stored in databases. Except for the authorized users, the end users are generally not required to access the database. Hence, it should be ensured the database servers are not publicly accessible unless the servers are providing hosting services. For organization's access to their servers, gateway servers can be set up for remote administrators (VPN or SSH tunnels).

4.14   Patching Database Servers

Database servers should be updated with current patches. Effective database patch management must be a proactive security approach because hackers continuously look for new database security gaps, and new viruses and malware are continually being produced. The timely deployment of the most recent service packs for the database servers, significant security hotfixes, and incremental upgrades can enhance the performance of the database servers.

4.15   Real-Time Database Monitoring

Enhancing security and enabling prompt response to future attacks are made possible by routinely reviewing your database for hacking attempts. Indeed, a File Integrity Monitoring (FIM) approach can assist in monitoring each activity performed on the database server. FIM can also alert database system users of potential security breaches. Ensure security teams are contacted whenever FIM discovers a modification to a vital database file so they can investigate and respond appropriately.

4.16   Web Application and Database Firewalls

To guard your database server against threats to database security, we should utilize a firewall. A firewall, by default, forbids access to traffic. Additionally, it should prevent our database from opening outbound connections, except if there is a specific need. We need to set up a web application firewall in addition to securing the database using a firewall (WAF). Because different attacks against web applications, such as SQL injection attacks, could be exploited to attain unauthorized entry to the databases. A database firewall may not block a majority of web application risks because standard firewalls work at the network layer. In contrast, web application layers work at the application layer, which is layer 7 of the OSI model. Therefore, a WAF can have the ability for detecting malicious activities of web application traffic. Examples of such application traffic are SQL injection attacks.

4.17 Database Auditing

Logging all database and file server activity is an additional security precaution. For security audits, login activity must be kept for at least a year. Any account that makes more unsuccessful login attempts than the allowed number ought to be immediately reported to the information security administrator for review. It's crucial to recognize changes to sensitive information and related permissions. You may first create efficient and accurate policies by using historical data to determine what data is sensitive, how it is being used, who is using it, and where it is going. You can also foresee how changes in your environment can affect security. You can discover previously unidentified threats by using this technique.

4.18 Vulnerability Assessments and Penetration Tests

Tools for vulnerability scanning, such as Nmap, OpenVas, and Nessus, are frequently used in vulnerability assessments. These utilities use an external machine to scan the environment for open ports and the services' version numbers. The administrator can confirm that the systems are following the endpoint security policies by comparing the test results with known services and patch levels that are supposed to be on the endpoint systems.

Penetration testing is checking a computer system, network, or online application for security holes that an attacker could exploit. Software programs can automate penetration testing, or they can be done manually. Penetration testing's primary goal is to locate security flaws. Penetration testing can be used to assess an organization's security policy's effectiveness, regulatory compliance, staff security awareness, and capacity to respond to and identify security issues. To guarantee consistent network security and IT management, organizations should conduct pen tests frequently-ideally, once a year.

5 Conclusion

This chapter aimed to discuss the vulnerability issues to the data and data repositories, such as data warehouses, data marts, data cubes, and data lakes that can store data in different formats, including structured, unstructured, or semi-structured. The chapter first introduced these data repositories. Then, vulnerability issues were discussed. Finally, the potential security measures were described to protect the data and its repositories.

References

1. Tubaishat A, Al Jouhi M (2020) Building a security framework for smart cities: a case study from uae. In: 2020 5th international conference on computer and communication systems (ICCCS 2020). IEEE, Electrical Network, 15–18 May 2020, pp 477–481
2. Rashid ANMB, Ahmed M, Pathan A-SK (2021) Infrequent pattern detection for reliable network traffic analysis using robust evolutionary computation. Sensors 21(9)
3. Rashid ANM, Ahmed M, Islam SR (2021) A supervised rare anomaly detection technique via cooperative co-evolution-based feature selection using benchmark unsw_nb15 dataset. In: Inernational conference on ubiquitous security. Springer, pp 279–291 (2021)
4. Rashid ANMB, Ahmed M, Sikos LF, Haskell-Dowland, P (2022) Anomaly detection in cybersecurity datasets via cooperative co-evolution-based feature selection. ACM Trans Manag Inf Syst 13(3)
5. Rashid ANMB, Choudhury T (2019) Knowledge management overview of feature selection problem in high-dimensional financial data: cooperative co-evolution and MapReduce perspectives. Probl Perspect Manag 17(4):340
6. Ahmed M, Mahmood AN, Islam MR (2016) A survey of anomaly detection techniques in financial domain. Future Gener Comput Syst 55:278–288
7. Ahmed M, Mahmood AN, Hu J (2016) A survey of network anomaly detection techniques. J Netw Comput Appl 60:19–31
8. Miloslavskaya N, Tolstoy A (2016) Big data, fast data and data lake concepts. Procedia Comput Sci 88:300–305
9. Saddad E, El-Bastawissy A, Mokhtar HMO, Hazman M (2020) Lake data warehouse architecture for big data solutions. Int J Adv Comput Sci Appl 11(8)
10. Rashid ANMB, Ahmed M, Ullah AB (2022) Data lakes: a panacea for big data problems, cyber safety issues, and enterprise security. In: Next-Generation enterprise security and governance, pp 135–162. CRC Press (2022)
11. Khine PP, Wang ZS (2018) Data lake: a new ideology in big data era. ITM Web Conf 17:03025
12. Shahabuddin M (2018) All about data repository. https://www.infoguardsecurity.com/all-about-data-repository/. Accessed 30 June 2022
13. Iosif A-C, Gasiba TE, Zhao T, Lechner U, Pinto-Albuquerque M (2022) A large-scale study on the security vulnerabilities of cloud deployments. In: Wang G, Choo KKR, Ko RKL, Xu Y, Crispo B (eds), Ubiquitous security, volume 1557 of Communications in computer and information science. 1st international conference on ubiquitous security (UbiSec). Guangzhou Univ, Guangzhou, Peoples R China, 28–31 Dec 2021, pp 171–188
14. All Blue Solutions (2022) Top 10 most common security vulnerabilities in database. https://www.allbluesolutions.com/blog/top-10-most-common-security-vulnerabilities-in-database/. Accessed 30 June 2022
15. Brook C (2018) What is a data repository? https://digitalguardian.com/blog/what-data-repository. Accessed 30 June 2022
16. Geekflare Editorial (2022) The most dangerous database threats and how to prevent them. https://geekflare.com/database-threats-and-prevention-tools/. Accessed 30 June 2022
17. Imperva (2021) Database security. https://www.imperva.com/learn/data-security/database-security/. Accessed 30 June 2022
18. Murphy D (2022) 10 ways to improve data security. https://www.lepide.com/blog/ten-ways-to-improve-data-security/. Accessed 30 June 2022

19. Sicilia M-A, Visvizi A (2019) Blockchain and OECD data repositories: opportunities and policymaking implications. Library Hi Tech 37(1, SI):30–42
20. Rashid ANMB, Ahmed M, Sikos LF, Haskell-Dowland P (2020) A novel penalty-based wrapper objective function for feature selection in Big Data using cooperative co-evolution. IEEE Access 8:150113–150129
21. Rashid ANMB, Ahmed M, Sikos LF, Haskell-Dowland P (2020) Cooperative co-evolution for feature selection in big data with random feature grouping. J Big Data 7(1):1–42
22. Techopedia (2022) Data cube. https://www.techopedia.com/definition/28530/data-cube/. Accessed 30 June 2022
23. Jiang Y, Jeusfeld M, Ding J (2021) Evaluating the data inconsistency of open-source vulnerability repositories. In: ARES 2021: 16th international conference on availability, reliability and security, 17–20 Aug 2021. Electrical Network

Misinformation Detection in Cyber Smart Cities

Anupom Mondol, Jeniya Sultana, Mohiuddin Ahmed,
and A. N. M. Bazlur Rashid

Abstract Since the modern city dwellers are highly adapting the cutting edge technologies to assuage their daily life problems as well as to optimize the management of city administration more efficaciously, collectively known as smart cities, which have copious amounts of misinformation escalated by the social media along with the IoT end-nodes tempered by human or errors in the process leads to a hostile a Cyber-Physical-Social System. Inevitably misinformation detection has been a very prominent research domain many researchers have already stepped into as it is clearly prophetic of creating chaos in city life. In the chapter, we provide cursory details about misinformation and its proliferating ways. Then, we present the impact of the misinformation in a smart city context. Afterward, we dive deeper into the detection techniques, which cover earlier analyses of state-of-the-art research.

1 Introduction

A cyber smart city can be defined as a technologically upgraded modern area that is provided with different types of electronic mechanisms, voice-controlled methods, and sensors to collect specific data. Regardless of the assistance of evolving cyber smart cities in various aspects like the lifestyle of citizens, businesses, and environment, those cities are prone to multiple threats for cyber-security, making it

A. Mondol
Bangladesh University of Textiles, Tejgaon I/A, Dhaka 1208, Bangladesh
e-mail: anupommondol@tmdm.butex.edu.bd

J. Sultana
Missouri State University, 901 S. National Ave, Springfield, MO 65897, USA
e-mail: js989s@MissouriState.edu

M. Ahmed · A. N. M. B. Rashid (✉)
Edith Cowan University, Joondalup, WA, Australia
e-mail: a.rashid@ecu.edu.au

M. Ahmed
e-mail: mohiuddin.ahmed@ecu.edu.au

© The Author(s), under exclusive license to Springer Nature Switzerland AG 2023
M. Ahmed and P. Haskell-Dowland (eds.), *Cybersecurity for Smart Cities*,
Advanced Sciences and Technologies for Security Applications,
https://doi.org/10.1007/978-3-031-24946-4_8

strenuous to fabricate a security maturity in them [16]. This chapter's motive and discussion is misinformation detection, an ongoing challenge because of day-to-day added complexities with advancing the latest technology in the cyber smart city.

However, several crucial components related to the smart city concept have already been implemented; worth mentioning can be firstly smart governance, secondly smart economy, thirdly smart people, fourthly smart mobility, fifthly smart infrastructure, and finally, smart living [16]. With the progress of technology, contemporary problems emerge. One of those is spreading misinformation on social media and data anomalies in IoT networks [3, 4, 22, 26, 27]. The exploding expansion of the world wide web has tremendously facilitated the process of people communicating with one another. Online social media, like Twitter, Facebook, Snapchat, YouTube, and Sina Weibo, have revolutionized information propagation, resulting in significant improvement in volume, velocity, and variety of online information transference. However, the divulgence of data is sped up by online social platforms, and also these media additionally accompanies the expansion of misinformation. As claimed by a survey conducted by Knight Foundation, Americans appraise that 65% of the news on social platforms is misinformation [8]. The circulation of fake news during a health crisis instigates trepidation among the people. For instance, news regarding COVID-19 from social media posts can include opinionated "natural remedies," "the origin of COVID-19", or about the vaccines and their side effects cascade to uncertainty in becoming immunized although vaccines are available.

Not only social platforms but also data-driven from devices, controllers, sensors, or connectors can be misleading if the system becomes malfunctions intentionally or unintentionally. This may cause great chaos when the centrally controlled system feeds in faulty data due to device or senor error, threat, attack, or malware.

The rest of the chapter is organized as follows: Sect. 2 presents the definition of misinformation. Section 3 discusses the impact of misinformation in a smart city context. Section 4 includes the misinformation detection techniques. Section 5 concludes the chapter.

2 Misinformation

In this era, the internet helps people spread numerous fallacious information on different digital communication platforms. Defining misinformation combines various forms of fake or inaccurate data that intentionally or unintentionally affect people's way of thinking.

In IoT networks, cyber misinformation covers two major types of fallacious information:

 i. Misinformation on digital media platforms.
 ii. Data anomaly form the end note of a Cyber Physical System.

Our primary focus lies on flawed information that has the embryonic ability to mislead people. So the formal definitions are discussed in the following section.

2.1 Definition and Classification of Misinformation

Information or a story that propagates across online social media and is sooner or later verified as false or inaccurate. Misinformation classification covers various types of imprecise information grouped by flawed intention, such as socio-economic advantages, political interference, deliberate trickery, or unintentional misrepresentation.

2.1.1 Fake News or Media Manipulation

Fake news or media manipulation has embellished the de-facto report for distinguishing erroneous information in various mainstream media. This category of false information was deliberately spread through media during turning events, such as in the 2016 U.S. Presidential Campaign.

Therefore, we can state that fake news is those news articles or stories that are intentionally formatted with a view to misleading readers or viewers. However, they can be verified and labeled as false by some other sources.

Our focus can be shifted to

- **Serious fabrications:** Serious fabrications can be defined as the prototypical formation of fake news, i.e., articles including a spiteful intent (e.g., pseudoscience writings and fake interviews), that very often become viral on social platforms.
- **Large scale hoaxes:** Records of fallacious information, which are camouflaged as genuine news, are known as hoaxes [29]. These hoaxes are generally fabricated on an immensely large scale compared to a simple news article with a view to picking out public figures or ideas.
- **Humorous fakes:** Humorous fakes are usually formatted to amuse viewers and readers where audiences are considered aware of the humorous content [5].

2.1.2 Rumors

Rumour can be defined as misinformation that formal origins have not yet been authenticated. Generally, rumors propagate by users on social media platforms.

However, at this point, we can claim that those interpretations hinge on the "unconfirmed" feature of the article, and this unjustified information can turn into authentic, partly authentic, completely false, or hover as unverified [40].

In the end, some researchers have classified rumors considering type, scope, and characteristics.

2.1.3 Data Anomaly

While people relish the universality and convenience of IoT networks and social media platforms, numerous malicious activities, e.g., bullying, terrorist attack organizing, and deceit data dissemination, can happen. Hence, it is tremendously predominant that we identify these abnormal activities within a very short time span in order to avert mishaps and attacks.

Formally speaking, data anomalies refer to inconsistencies in the information stored or something that is abnormal or does not fit in.

There are two major categories of an anomaly:

- Point Anomaly: The irregular behavior or action of individual user.
- Group Anomaly: The aberrant pattern conducted by group of people.

Data anomalies can be the product of system malfunction in the smart network system. When any device or connection of the IoT network transmits wrong information intentionally or unintentionally, that piece of data may cause serious faulty fabrication to the overall data storage.

Aside from fake news and rumors, there are other forms of misinformation on social media.

3 Impact of Misinformation in a Smart City Context

The world population and the government want to lead more privileged daily lives where they can get maximum utilization of time and money through the cutting-edge technology where the concepts and implementation of the smart city showed up. In this domain of newly popped-up technologies, a small, vulnerable task can create compromised cyber security, leading to a disastrous cyber-physical system by one individual or an organization. Amongst all the reasons behind a compromised cyber-physical social system, misinformation is one of the deadliest possible impacts of this modern society or the smart city. Misinformation can be formed as advertising deceptively, forging a document, deepfaking images and videos, fabricating history, or manipulating entries into some trusted website like Wikipedia [10]. The components of a smart city are connected: physical infrastructure, social elements, business infrastructure, and information technology computing. Misinformation challenges all of these four basic components of a smart city. We will elaborate on the impacts to the point of how the misinformation affects city security, social disruption, and shaping public opinion and political situation.

3.1 Impact on Cyber Security

Although the advancement of information and technology, modern city dwellers have leveraged their daily life by consuming less energy, having smart time management, indulging in social connections, and being vibrant. Cyber security undermined by spreading misinformation is a big issue to be solved and taken care of daily by the concerned authorities, including researchers, governing bodies, and technology providers. Cyber-security covers every internet-connected smart device from malicious attacks by preserving the privacy of used devices, safeguarding the information of the devices, and protecting the information using people's identity [36].

Every new solution for city dwellers in a smart city creates a new field of cyber-attacking. For example, the cyber intelligent system of traffic signaling light use no encrypted transmission making it attractive to a cyber attacker to manipulate data making the system so vulnerable. On the other hand, misinformation can automatically be generated from many end-nodes of a cyber-physical system since a cyber-physical system generally implemented on IoT has many sensors at the end of the network to opt for collecting data from the physical world. Figure 1 shows the impact of misinformation in the smart city on how it spreads and what results in different sectors.

Smart building is one of the most salient elements of a smart city. It uses sensors and smart grid technologies for interconnecting and transmitting data between building components. It afterward sends to a smart meter the amount of energy consumption. The whole system automatically adjusts the energy requirements, consumption, and profiles along with controlling the electrical elements of the building remotely[15, 17]. This system has a huge scope to misinterpret the information of user sensitive data can be revealed such as erroneous data of power consumption as well as the type of electrical home appliances, or there is any member in the house of not [14]. In addition, eavesdropping by bad people like a thief or robbers may get protected information and damage their privacy [38]. A smart grid is highly

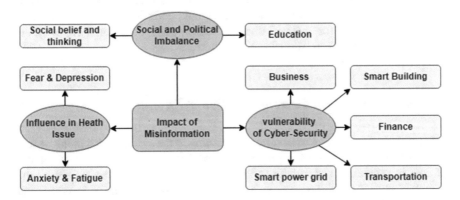

Fig. 1 Impact of misinformation in smart city

dependent on communicating amongst various nodes of a smart city, and the grid component makes it the most vulnerable system. Misinformation can be put from any of the nodes of this smart grid system.

Misinformation can bring a misleading future to the smart transportation system in smart cities since the attacker. It can be done by sending misinformation like fake alerts, false locations, or forgery of promotional data, which could mislead drivers [37].

3.2 Influence on Health System

In a cyber smart city, misinformation causes serious damage to public health, and we will analyze the covid pandemic in the following context. When all county was going on a locked-down situation, digital media and social platforms were the only way to inform us about the virus's transmission, clinical manifestation, prevention of the virus, and death counts caused by the virus. This information generally creates panic causing physical and psychological disorders and comprising the population's immune system [35]. This emergence of the pandemic put society into health disorders including anxiety, sleeping disorder, depression, and even increasing in suicidal death [6]. Even the nurse is challenged with the mantel health [20].

4 Misinformation Detection Techniques

As we mentioned earlier, misinformation can be generated from any node in the cyber-physical system of a smart city. This is because it is a complex network that adopts the Internet of Things (IoT), making interactions with intelligent physical and computational components with infinite data generative nodes. On the other hand, misinformation like fake news, fake advertisement, and fake social and political issues can spread over the mass connecting digital platform known as social media of smart cities. Therefore we will divide misinformation detection into the cyber-physical system (CPS) and digital social platforms.

4.1 Misinformation Detection in Cyber-Physical System

The revolutionary concept of smart city is coming to the existent for helping people with the numerous scopes of leveraging ways of daily life in which copious amount of data is generated, and infinite actions have to be taken in each second by the computer or city brain. So smart cities need cloud storage connected to the many networks built with IoT sensors at the end to communicate with the physical world.

Generally, smart city uses fog computing architecture with three layers: cloud, fog, and sensors. The fog layer is used between cloud and IoT sensors layers because of overcoming latency issues [30].

The network-based intrusion detection works on network data traffic, which has two approaches: signature-based and anomaly-based. Among them, anomaly-based detection has been more efficient for identifying new attacks than others and has more false positive detection. There are many techniques for attacking a cyber-physical system, such as brute-force attack, Buffer overload attack, Phishing, attack by SQL injection, Sniffer attack, Trojan horses, and many more.

There are many machine learning and deep learning algorithm to classify suspicious behavior by inspecting the misinformation put through in the network by the intruder. These algorithms can identify whether an intruder spreads misinformation on the line or not by interpreting anomalies in the data.

4.1.1 Machine Learning Algorithm for Detection

K-Means

This is an unsupervised interactive learning method that separates a data set into several district groups which are nonoverlapping. This number of distinct groups can be set previously; this is called K. It classifies data into a subgroup where the distance after squaring between all data points and the centroid of that class is at the lowest. It gives us the assumption about the spherical Gaussian distribution of data clusters with various means with the same covariance matrix.

Naive Bayes Classifier

Bayes' theorem is a basic rule probability, which gives the probability that an event will happen in the presence of another event happening. A frequency table is built for each attribute against a target class, which is transformed into another table and passed through Eq. 1 to deduce the probability of each class. The prediction results in the class with the largest probability, and the target is to predict the right class for new data.

K-Nearest Neighbors Classifier

As a supervised learning algorithm, K nearest neighbors (kNN) depend on the Euclidean distance between two data. Suppose this algorithm classifies a given data y. In that case, firstly, this algorithm deduces all the Euclidean distances from the given data of all groups' points and sorts them out in ascending order. Then it will be distinct the top k data and classify the y data to the majority of the k nearest data.

DBSCAN

DBSCAN stands for Density-Based Spacial Clustering of Application with Noise which works with a density of data and cluster data, discarding noisy data. It also uses Euclidean distance. So it is not generally used for sparse data with high dimensional data, unlike k means clustering. DBSCAN uses two arguments to cluster all the points, including the radius of neighbor data points and a minimum number of data points within that radius. The distance metric calculates the radius of the data points in a neighborhood, and a circular area of this radius centering a core point covers the minimum data points. Therefore, the data outside the core points' radius are considered noisy data.DBSCAN is efficient for the cyber smart city context because it denies noise processing. The most significant advantage of this clustering algorithm is that the parameter of the minimum point and the radius can be set by the engineer or expert who can understand and predict the accepted range of data. So it can detect anomalies in the data packet and cut them out.

Decision Trees

It is a supervised learning algorithm used for classification. It looks like a tree topped with a root node containing all the data set, intermediate nodes, and branches representing the decision made by the sub-tree, and the leaf nodes containing a distinct attribute or class of data. First, it searches for the best attribute in the data set contained by the root through the Attribute Selection Measure (ASM). Then, this data set will be divided into subsets according to the best attribute, which leads to the construction of an internal node. Repeating these actions of dividing into a subset and searching for the best attribute leads to data fitting with the best attribute and classification. Selecting the best attribute through ASM is a key topic in the decision tree. There are some works on intrusion detection with the CART decision tree, and [21] is one of them.

SVM

Support Vector Machine (SVM) is a very popular supervised learning technique for binary classification problems. It creates a linear or non-linear line between the two classes of data. This n-dimensional best line is known as a hyperplane. In another way, a threshold creates two separate groups of data. For instance, a dog and cat classifier is popular on the internet nowadays and has used the SVM method to classify whether it is a dog or a cat. Now, if we give input to the classifier a strange cat that has some dog properties, the support vector can classify it as a cat because of the hyperplane. In the intrusion detection system, SVM is used to classify the good data from the threat data. Soft margin and kernel tricks are the two techniques introduced by the SVM, and the first one is very good at maximization of distance amongst classes of important and intrusion data.

Isolation Forest

Isolation forest is a great tool for anomaly detection in data packets being passed through the smart network and is used widely in developing smart cities. It implies binary trees to separate misinforming data with a low constant linear complexity and need lower space to process significant amounts of data. It breaks the data space apart using an orthogonal line against the origin, giving an anomaly sore to the data points and requiring a few splits for segregation. In the training phase of this detection algorithm, an isolation tree is built. After picking a subset from the data, an attribute and a data point of this attribute are randomly chosen between the minimum and the maximum value of the subset, called a split value. Then, the left branch of the isolation tree expands with a chosen data point if it cannot outnumber the split value, and in contrast, the right branch of the isolation tree is constructed. This step occurred for every data point until a preset depth of this tree was achieved, and this was the training phase of the isolation tree. After this phase, acquire the path length of every data point. If we want to detect data anomalies, we need to calculate the anomaly score for the data point. To calculate the anomaly score for data, x, we need to do the followings.

$$s(x, m) = 2^{\frac{-E(h(x))}{c(m)}} \tag{1}$$

Where h(x) is path length of data x, E(h(x)) is average of h(x) from the isolation tree, s(m, x) is the anomaly score of x with data subset size of m and

$$c(m) = 2H(m - 1) - \frac{2(m - 1)}{n} \tag{2}$$

where the size of testing data is n and H is a harmonic number which can be represented by $H(i) = ln(i) + \gamma$ where γ is the Euler-Mascheroni constant. When m is equal to 2, the value of c(m) is 1, and if m is less than 2, c(m) is 0.

There are some interesting findings to detect anomalies from the anomaly score of data. The data x is an anomaly if the anomaly score s(m, x) is nearly 1, and a normal value if the anomaly score is below 0.5. Therefore the data set is called anomaly free if the anomaly score of all data points of a data set is nearly 0.5.

The advantage of using isolation forests is very significant in the cyber smart city context because the smart city produces big data every day and anomalies in these data remain scattered and unstructured patterns [28].

Gradient Boosting Machine

Gradient Boosting Machine (GBM) based classification has been used in data anomaly detection for a long run, especially where the ratio of anomaly and actual data is huge. This machine learning algorithm does classification and regression simultaneously and improves classification and regression trees. In this method, a tree is generated and expanded consecutively depending on the previous tree. The culminating prediction or detection can be achieved by adding up the leaves' weight after turning a sample data to the leaves of the tree by the decision rules. Gradient boosting is more accurate than SVM and DNN in anomaly detection [34].

4.2 Misinformation Detection by Natural Language Processing

Advanced Natural Language Processing, known as NLP algorithms, has been a prominent tool for the featurization of the text article supporting the detection process of misinformation. Since it is a subdomain of Artificial Intelligence (AI) uses some popular machine learning techniques for NLP in misinformation detection, including RNN, LSTM, BiLSTM, Naive bais classifier, Basic logistic regression with binary classification models, CNN, SVM, and Decision tree-like Random Forrest Classifier (RFC). In each of these processes, the input article, a long string, is tokenized or split into words and then stems along with weighting them. Then, features are extracted from the tokenized words via many methods like TF-IDF (Term Frequency-Inverse Document Frequency), which is responsible for finding the relevancy of words in an article, and LIWC (Linguistic Inquiry and Word Count), which is for counting and analyzing linguistic features. After these data preprocessing steps, those classification learning algorithms are used to detect textual misinformation.

4.2.1 Deep Learning Algorithm for Detection

Autoencoders

Autoencoders are unsupervised and very simple deep learning algorithms that are very useful for reduction techniques. These algorithms encode the input, store it as a code and try to decode it as it was since it has three-layer including encoder, code, and decoder. The output is similar to the input with degraded dimension and compressed. It can be called a self-supervised learning algorithm since it creates its levels during the time of training. It is widely used for anomaly detection. We know the output data is reconstructed from the code, which is encoded input. This process can produce reconstruction errors which are pondered as anomaly scores. Adding up the mean and the standard deviation of these anomaly scores obtained from reconstruction errors produces a threshold. The reconstruction error of data above this threshold is considered an anomaly.

Recurrent Neural Network

Amongst deep learning algorithm Recurrent Neural Network (RNN) is fit most for classifying sequential data since it evaluates the time of the data. RNN differs from ANN with a hidden layer that contains a vector state of prior inputs or outputs. An RNN is even used with convolutional layers to elaborate the active pixel in the neighborhood. The equation and equation can represent a simple mathematical representation of RNN.

$$h_t = \sigma^h (W^h h_{t-1} + W^x x_t) \tag{3}$$

$$y_t = \sigma^y (W^y h_t) \tag{4}$$

Where x_t is data sample at the time t, h_t is current and h_{t-1} is the previous hidden state vector. W_y, W_x and W_h are the dense metrics and σ^h and σ^y. are the activation operator.

LSTM

Long Short Term Memory algorithm (LSTM) is a kind of RNN that is the most used deep learning method in the 20th century. Input and output gates, a cell state, and a forget gate are the portions of a classic RNN model. The forget gate is used to keep data from the previous operation. The LSTM differs from RNN in the ability to keep previous data or not and a cell state which contains the previous long-term data. The forget gate decides whether the data should be remembered or not. Along with LSTM, BiLSTM and Gated Recurrent Unit (GRU) techniques are also being used.

One Dimensional Convolutional Neural Networks (CNN)

Generally, 2D CNN is the best recognized and used for image processing, but in anomaly detection, 1D CNN is significantly sound. The amount of hidden CNN and MLP layers, kernel size, sun-sampling factor, and the chosen pooling and activation function are the parameter before working with 1D-CNN.

4.3 Multi Modal Misinformation Detection in Cyber Digital Platform

Misinformation spreaders are used to spread misinformation via social media in text, images, audio, video data, or other modalities. However, the scenario is different since misinformation spreaders collaborate amongst the forms of spreading misinformation, such as blending text and image or video and audio. Nowadays, visual information is being very commended to the people, and spreaders are taking advantage of it. Thus, misinformation detection researchers have been constructing new methods to detect this cross-field discordance in the digital platform. There are many intermodal dissimilarities, including:

- Contradicting text and image
- Contradiction between video and description style
- discordance between domain and propagation Network of Text on web
- All inclusive look of a domain
- Manipulation of Audio and video
- Deepfaking audio and video.

Amongst these multi-modal dissimilarities of data, deepfake is the most cutting edge and venomous of all. Deep-fake is a deep learning and faking technique to vigorously manipulate an existing image, video, and audio where a person's face, voice, or other attributes are replaced with others. It is used as a lethal weapon in blackmailing, politics, art, acting, and creating a meme. In social media, even deepfake creates a non-existing person who is active on social media or traditional media, such as Oliver Taylor [31]. Deepfake is very brutal in spreading misinformation.

The research is going on to detect deepfake in many universities with an accuracy rate above 99% sometimes. The most well-known technique to detect the deepfake photo or video is the same as how it was built [18]. A team of researchers used only light reflection to detect deepfake with high accuracy published on 2020 [32]. In terms of misinformation detection in multi-dimensional mode, researchers invested themselves in finding many solutions, including machine learning and deep learning techniques.

Classic Machine Learning

Many machine learning techniques benefit from fake news detection using the text of news articles. There are also some works to detect misinformation based on images only [1]. These techniques lift the single modality feature. However, there is not much research on classifying multi-modal aspects of news articles using machine learning techniques. Some researchers encode article text, social content contexts, and host domain features into segregated embeddings and construct a shared structure, which is used to classify the article [2, 23–25].

Deep Learning Solutions

In the arena of misinformation detection, deep learning methods have been the most useful tools to classify video, images, text, and some other modalities for the past few years. These deep learning methods can be divided into five parts as followings with some details.

Concatenation-Based Methods

The lion's share method for misinformation detection is to set individual embeddings of image, text, or other modalities into a vector and combine them to get a multi-modal vector representation used for classification. Researchers suggested implementing XLNet[1] which is an auto-regressive model, and Vgg-19[2] which is 19 layers deep CNN, on text and image respectively and then combining the feature vector leads to classify and detect the misinformation [19].

Attention-Based Methods

Concatenating the feature vectors in every misinformation detection may fail to build multi-modal embeddings since the whole text of an article may not be incorrect for any given image, or the image may not be fabricated against the text to declare it as misinformation. Therefore the attention-based method is introduced to solve this problem which is more effective in exploiting embeddings with higher multi-modalities. TRANSFAKE is a method that constructs a series with the attributes of text and images and puts them through a vision-language transformer (ViLT) [13] model to know the multi-modal properties [12].

Generative Methods

Many single and multi-modal misinformation classification techniques utilize generative learning methods that can disperse data by themselves and predict or generate the next data pattern. These methods augment the accuracy of the detection classifier by far. For instance, a generative algorithm has been implemented on a BERT-based Variational autoencoder, which has an encoding technique, a decoding technique, and a binary classifier to detect misinformation. The shared and multi-modal representation of visual and textual information is encoded to a multi-layered latent vector, which is then decoded back to the initial image and text. Later this shared information is fed to the binary classifier to identify the fake article [11].

Graph Neural Network Method

In recent years, graph Neural Network (GNN) based detection techniques have been used to classify misinformation [7, 33]. In this technique, all the media contents like text and images are converted to a graph, and GNN draws out the semantics of misinformation afterward. For another example of using GNN-based misinformation detection, we can recall the research in which they built a graph for each social

---

[1] https://github.com/zihangdai/xlnet.

[2] https://keras.io/api/applications/vgg/.

content depending on the point-wise information (PMI) score of a word couple and brought out the required information through knowledge distillation [9].

Multi Modal Discordance-Aware Methods
The previously stated deep learning models implemented on different modalities show us that fabricating in any modalities leads to mismatching representation and falling into misleading and misinterpreted news articles. So, there is a cross-modal dissonance to be detected by the customized method. Therefore resemblance of textual information with the other modalities like visual and sound information is a cue for assuming news content is vulnerable. SAFE is a similarity-based multi-modal misinformation detection method by deducting relevance between text and other modalities like images and videos by an adaptive cosine similarity [39].

5   Conclusion

Human society has been evolving from the beginning of humankind, depending on the availability of living utility. The modern world surrounded by technologies offers us to live in a tranquil standard of livelihood where smart cities appear but with so many challenges like cyber misinformation. In this chapter, misinformation in the cyber smart city is elaborated on with examples and how it spreads. Primarily we tried to represent the current research on misinformation detection in a smart city context along with the impact of the misinformation on an IoT-based cyber-physical system. An attacker can generate misinformation from the IoT-based cyber-physical network, or the misinformation spreader can originate on digital communication platforms. The misinformation detection technique is categorized by the modality of the data domain, along with cutting-edge research. The research gap was explained later on as well as proposing some ideas to support the misinformation detection on which fruitful and copious research could be done, such as joining the existing research algorithm with the cyber attack learning algorithm.

References

1. Abdali S, Gurav R, Menon S, Fonseca D, Entezari N, Shah N, Papalexakis EE (2021) Identifying misinformation from website screenshots. In: Proceedings of the International AAAI Conference on Web and Social Media 15(1):2–13
2. Abdali S, Shah N, Papalexakis EE (2020) Hijod: semi-supervised multi-aspect detection of misinformation using hierarchical joint decomposition. arXiv:2005.04310
3. Ahmed M, Mahmood AN, Hu J (2016) A survey of network anomaly detection techniques. J Netw Comput Appl 60:19–31
4. Ahmed M, Mahmood AN, Islam MR (2016) A survey of anomaly detection techniques in financial domain. Futur Gener Comput Syst 55:278–288
5. Bondielli A, Marcelloni F (2019) A survey on fake news and rumour detection techniques. Inf Sci 497:38–55

6. Choi EPH, Hui BPH, Wan EYF (2020) Depression and anxiety in Hong Kong during covid-19. Int J Environ Res Public Health 17(10):3740
7. Cui L, Seo H, Tabar M, Ma F, Wang S, Lee D (2020) Deterrent: Knowledge guided graph attention network for detecting healthcare misinformation. In: Proceedings of the 26th ACM SIGKDD international conference on knowledge discovery & data mining, pp 492–502
8. Guo B, Ding Y, Yao L, Liang Y, Yu Z (2019) The future of misinformation detection: new perspectives and trends. arXiv:1909.03654
9. Hinton G, Vinyals O, Dean J (2015) Distilling the knowledge in a neural network
10. itcs.ecu.edu. Fake News, Misinformation, and Disinformation as Cyber Security Threats (2021). https://itcs.ecu.edu/itcs-news/fake-news-misinformation-and-disinformation-as-cyber-security-threats/, MAY 21, 2022. Last accessed on July 21, 2022
11. Jaiswal R, Singh UP, Singh KP (2021) Fake news detection using bert-vgg19 multimodal variational autoencoder. In: 2021 IEEE 8th Uttar Pradesh section international conference on electrical, electronics and computer engineering (UPCON). IEEE, pp 1–5
12. Jing Q, Yao D, Fan X, Wang B, Tan H, Bu X, Bi J (2021) Transfake: multi-task transformer for multimodal enhanced fake news detection. In: 2021 international joint conference on neural networks (IJCNN). IEEE, pp 1–8
13. Kim W, Son B, Kim I (2021) ViLT: vision-and-language transformer without convolution or region supervision. https://arxiv.org/abs/2102.03334. Last accessed on July 22, 2022
14. Kumar A, Sharma S, Goyal N, Singh A, Cheng X, Singh P (2021) Secure and energy-efficient smart building architecture with emerging technology IoT. Comput Commun 176:207–217
15. Lyu C, Jia Y, Zhao X (2021) Fully decentralized peer-to-peer energy sharing framework for smart buildings with local battery system and aggregated electric vehicles. Appl Energy 299:117243
16. Ma C (2021) Smart city and cyber-security; technologies used, leading challenges and future recommendations. Energy Rep 7:7999–8012
17. Mehrpooya M, Ghadimi N, Marefati M, Ghorbanian SA (2021) Numerical investigation of a new combined energy system includes parabolic dish solar collector, stirling engine and thermoelectric device. Int. J. Energy Res. 45(11):16436–16455
18. Newsgaurd (2022) Combating misinformation with trust ratings for news. https://www.newsguardtech.com/. Last accessed on July 22, 2022
19. Nie H, Han X, He B, Sun L, Chen B, Zhang W, Wu S, Kong H (2019) Deep sequence-to-sequence entity matching for heterogeneous entity resolution. In: Proceedings of the 28th ACM international conference on information and knowledge management, pp 629–638
20. Okechukwu EC, Tibaldi L, La Torre G (2020) The impact of covid-19 pandemic on mental health of nurses. La Clinica Terapeutica 171(5)
21. Radoglou-Grammatikis PI, Sarigiannidis PG (2018) An anomaly-based intrusion detection system for the smart grid based on cart decision tree. In: 2018 global information infrastructure and networking symposium (GIIS). IEEE, pp 1–5
22. Rashid ANM, Ahmed M, Islam SR (2021) A supervised rare anomaly detection technique via cooperative co-evolution-based feature selection using benchmark unsw_nb15 dataset. In: International conference on ubiquitous security. Springer, pp 279–291
23. Rashid ANMB, Ahmed M, Sikos LF, Haskell-Dowland P (2020) Cooperative co-evolution for feature selection in big data with random feature grouping. J Big Data 7(1):1–42
24. Rashid ANMB, Ahmed M, Sikos LF, Haskell-Dowland P (2020) A novel penalty-based wrapper objective function for feature selection in Big Data using cooperative co-evolution. IEEE Access 8:150113–150129
25. Rashid ANMB, Choudhury T (2019) Knowledge management overview of feature selection problem in high-dimensional financial data: Cooperative co-evolution and MapReduce perspectives. Probl Perspect Manag 17(4):340
26. Rashid ANMB, Ahmed M, Pathan ASK (2021) Infrequent pattern detection for reliable network traffic analysis using robust evolutionary computation. Sensors 21(9)
27. Rashid ANMB, Ahmed M, Sikos LF, Haskell-Dowland P (2022) Anomaly detection in cyber-security datasets via cooperative co-evolution-based feature selection. ACM Trans Manag Inf Syst 13(3)

28. Rashid ANMB, Ahmed M, Ullah AB (2022) Data lakes: a panacea for big data problems, cyber safety issues, and enterprise security. In: Next-generation enterprise security and governance. CRC Press, pp 135–162
29. Rubin VL, Chen Y, Conroy NK (2015) Deception detection for news: three types of fakes. Proc Assoc Inf Sci Technol 52(1):1–4
30. Santos J, Leroux P, Wauters T, Volckaert B, De Turck F (2018) Anomaly detection for smart city applications over 5g low power wide area networks. In: NOMS 2018-2018 IEEE/IFIP network operations and management symposium. IEEE, pp 1–9
31. Satter R (2022) Deepfake used to attack activist couple shows new disinformation frontier. https://www.reuters.com/article/us-cyber-deepfake-activist-idUSKCN24G15E, JULY 15, 2020. Last accessed on July 21, 2022
32. security.org. Misinformation and disinformation: A guide for protecting yourself. https://www.security.org/digital-security/misinformation/#facts, August 16, 2021. Last accessed on July 22, 2022
33. Song C, Shu K, Bin W (2021) Temporally evolving graph neural network for fake news detection. Inf Process Manag 58(6):102712
34. Tama BT, Rhee K-Y (2019) An in-depth experimental study of anomaly detection using gradient boosted machine. Neural Comput Appl 31(4):955–965
35. Tangcharoensathien V, Calleja N, Nguyen T, Purnat T, D'Agostino M, Garcia-Saiso S, Landry M, Rashidian A, Hamilton C, AbdAllah A et al (2020) Framework for managing the covid-19 infodemic: methods and results of an online, crowdsourced who technical consultation. J Med Internet Res 22(6):e19659
36. Uchendu B, Nurse JRC, Bada M, Furnell S (2021) Developing a cyber security culture: current practices and future needs. Comput Secur 109:102387
37. Xie S, Zhijian H, Wang J, Chen Y (2020) The optimal planning of smart multi-energy systems incorporating transportation, natural gas and active distribution networks. Appl Energy 269:115006
38. Xie X, Lu Q, Herrera M, Yu Q, Parlikad AK, Schooling JM (2021) Does historical data still count? exploring the applicability of smart building applications in the post-pandemic period. Sustain Cities Soci 69:102804
39. Zhou X, Wu J, Zafarani R (2020) Similarity-aware multi-modal fake news detection. In: Pacific-Asia conference on knowledge discovery and data mining. Springer, pp 354–367
40. Zubiaga A, Aker A, Bontcheva K, Liakata M, Procter R (2018) Detection and resolution of rumours in social media: a survey. ACM Comput Surv (CSUR) 51(2):1–36

Vehicle Trajectory Obfuscation and Detection

Baihe Ma, Yueyao Zhao, Xu Wang, Zhihong Liu, Xiaojie Lin, Ziwen Wang, Wei Ni, and Ren Ping Liu

Abstract A vehicle in road networks shares location data with other vehicles and location-based services (LBS) through Internet-of-vehicles (IoV). By analyzing the location data from vehicles, LBS providers can offer vehicles better services. However, fake trajectories created by adversaries and malicious drivers diminish the location data utility in IoV and breach the quality of LBS. Illegal trajectory detection is vital to ensure location data utility in IoV. Existing location privacy-preserving schemes like obfuscation schemes add noise to actual location data increasing difficulties in detecting illegal trajectories. In this paper, we detect illegal trajectory in the case that all drivers in road networks protect location privacy by using obfuscation. We propose a new personalized obfuscation mechanism to dynamically and adaptively protect the location privacy of drivers in road networks. Considering uneven

B. Ma (✉) · X. Wang · X. Lin · R. P. Liu
Global Big Data Technologies Centre, University of Technology Sydney, Ultimo, NSW 2007, Australia
e-mail: Baihe.Ma@uts.edu.au

X. Wang
e-mail: Xu.Wang-1@uts.edu.au

X. Lin
e-mail: Xiaojie.Lin@uts.edu.au

R. P. Liu
e-mail: RenPing.Liu@uts.edu.au

Y. Zhao · Z. Liu · Z. Wang
Xidian University, Xian, China
e-mail: liuzhihong@mail.xidian.edu.cn

Z. Wang
e-mail: wangziwen@stu.xidian.edu.cn

W. Ni
Data61, CSIRO, Sydney 2122, Australia
e-mail: Wei.Ni@data61.csiro.au

© The Author(s), under exclusive license to Springer Nature Switzerland AG 2023
M. Ahmed and P. Haskell-Dowland (eds.), *Cybersecurity for Smart Cities*,
Advanced Sciences and Technologies for Security Applications,
https://doi.org/10.1007/978-3-031-24946-4_9

protection, we propose a trajectory detection scheme to classify trajectories in IoV. We evaluate our detection method with the data of real-world road networks, which is an important scenario of smart cities. The experiment results show that the proposed classifier outperforms existing studies in detecting malicious obfuscated trajectories with at least 94% of the Area Under the Curve (AUC) score.

1 Introduction

Location-based services (LBS) have been extensively and deeply developed as an important part of smart cities to provide various services, e.g., traffic analysing and urban planning [1, 2]. Smart cities inevitably require spatio-temporal location data of vehicles, which is highly correlated with a driver's private information (e.g., home address, company address, and religion) [3]. Location-based services (LBS) providers rely on location data from drivers to offer various services [4]. However, adversaries can infer drivers' personal information by analyzing location data [5]. It is of necessity to protect location data in Internet-of-vehicles (IoV) for drivers' privacy while ensuring high quality of service (QoS).

Obfuscation schemes [6] and adaptive location privacy-preserving schemes [7] have been developed to protect location privacy. Obfuscation schemes add noise to obfuscate drivers' actual location data, which reduces the data utility of the location data. The obfuscated location data are indistinguishable from each other, which leads to the fact that LBS providers would collect illegal location data. The location privacy-preserving schemes aim to balance the QoS of LBS and privacy-preserving capability by analyzing the drivers' requirements.

Malicious drivers breach legal drivers' profit. For example, malicious drivers can occupy more benefits than they deserve by deliberately modifying their trajectory data in Taxi service [8, 9]. Malicious drivers also use location privacy-preserving schemes to protect their location data. By analyzing location data which concludes illegal location data, smart city applications cannot provide an acceptable QoS of LBS. Therefore, LBS should detect the illegal location data to ensure high QoS. If the malicious drivers employ location privacy-preserving schemes (e.g., obfuscation schemes) as the legal drivers, detecting illegal data becomes difficult.

We study the illegal location data detection and propose a personalized obfuscation scheme and an illegal trajectory detection mechanism. Our work has two-fold contributions as follows:

- We employ the differential privacy in the proposed personalized obfuscation scheme to protect drivers' location privacy adaptively and to provide high QoS of LBS in road networks.
- We propose a Convolutional Neural Network (CNN) based detection mechanism to detect illegal trajectories without requiring the drivers' actual location. The proposed scheme has high accuracy in detecting illegal trajectories even if the drivers protect actual location data with various noise sizes.

We conduct experiments with the real-world road network dataset extracted from Open Street Map (OSM)[1] to evaluate the proposed scheme. The illegal trajectories are generated with fake speeds and paths with the real-world road network dataset. We also evaluate the proposed scheme in the case that the drivers adaptively protect their location data, i.e., using different privacy levels. The experimental results show that the proposed detection scheme achieves at least 94% Area Under the Curve (AUC) score when detects the illegal obfuscated trajectory.

The rest of this paper is organized as follows. Section 2 studies the existing works. Section 3 describes the proposed mechanisms. Section 4 evaluates the proposed mechanisms with real-world road network dataset. Section 5 concludes the paper.

2 Related Works

The existing smart city provides location-based services by mining trajectory data that are transmitted in vehicular networks [10]. Wang et al. have studied the privacy challenges in smart city and analyzed the privacy leakage in LBS [11]. The authors pointed out that the smart city can provide high quality of services if trajectory privacy is well protected.

The previous studies of location obfuscation mechanisms perturb a driver's actual location and report an obfuscated version to LBS. Derived from the differential privacy [12], scheme in [6] first developed the concept of geo-indistinguishability. The scheme follows the idea of geo-indistinguishability and uses Laplace distribution to add controlled noise for protecting location data locally. Yu et al. [13] improved the two-phase dynamic differential location privacy scheme by integrating the inference error expectation and geo-indistinguishability [14]. The improved framework effectively protects location privacy in a 2D map. The authors developed an adaptive location privacy-preserving mechanism in [7] to balance location privacy and utility. The mechanism calculates the amount of noise before adding noise to actual location data. The calculation is based on the correlation level between the driver's current location and the previous obfuscated locations. With the concept of differential privacy, Xiao et al. [15] improved a location-cloaking system to protect drivers' location data in a 2D map. The obfuscation locations generated by the existing 2D obfuscation mechanisms might locate at unreachable locations, e.g., in the river, which breaches the location privacy-preserving capability of the obfuscation mechanisms.

The existing illegal trajectory detection mechanisms are classified into the machine-learning-based detection and the rule-based detection. The illegal trajectory data is detected by utilizing GPS data in the rule-based detection mechanisms. Machine-learning-based detection mechanisms classify trajectory data as legal and illegal by using techniques like deep neural networks. In [16], Chen et al. improved

---

[1] Open Street Map is an open source database of the world's geographic map. https://www.openstreetmap.org/.

an efficient real-time trajectory detection method with low processing overhead. The method uses the window size to estimate the partial trajectory that result in the anomalousness trajectory. The trajectory detection system with two-phase outliers upon trajectory data streams was improved in [17]. The two phases are the trajectory simplification and the outlier detection.

In [18], authors developed a trajectory detection method based on a recurrent neural network (RNN). The authors extracted drivers' behaviors within a sliding window and uses the deep representations that are fixed-length for the feature sequence. The authors grouped the representations into clusters before detection. CNN is first introduced in [19], which is further utilized in the fields such as the natural language processing and speech recognition. A CNN-based trajectory prediction method was improved in [20]. The method simplifies the network structure and utilizes the trajectory structure (spatio-temporal consistency). The experimental results show that the CNN-based trajectory prediction method can detect illegal trajectories with a high score of the AUC.

Our work obfuscates drivers' trajectories in the road networks to avoid the generation of the off-road obfuscated locations. Then, an illegal trajectory detection scheme based on CNN is proposed in this paper. The proposed scheme does not expose the drivers' actual trajectories and achieves high detection accuracy. The proposed scheme can detect illegal trajectories even if the drivers use various privacy parameters to obfuscate the locations. To the best of our knowledge, we are the first work that detects illegal locations in real road networks, which is almost the actual usage scenario of a smart city.

3 Proposed Scheme

In this paper, we start by proposing an adaptive obfuscation scheme to customized protect location privacy in road networks. Then, we propose an illegal trajectory detection system with CNN to identify the legal and illegal trajectory from the obfuscated trajectory. In our model, X_m is the mth trajectory consists of the location points sequence $(x_{m1}, x_{m2}, \ldots, x_{mn})$. $x_{mi} = (lat_i, lon_i)$ is a tuple which stands for the coordinate (i.e., latitude lat_i and longitude lon_i) of a location.

3.1 System Model

The existing obfuscation studies [7, 21, 22] pay attention to protect the drivers' location data in a 2D map, which generate off-road locations (e.g., railroads and rivers). Adversaries can exclude the off-road obfuscated locations from real trajectory data when they identify the obfuscated locations that are off-road. The Euclidean distance between the obtained off-road location and the nearby road can be utilized by the adversaries to estimate the actual location. In this paper, we controlled the obfuscated candidates to avoid the off-road data and guarantee that all obfuscated locations are on-road.

3.2 Dynamic Obfuscation Scheme

Definition 1 (*Geo-indistinguishability* [6]) Let P be a probabilistic function. Let X and Z be a set of the actual location candidates and obfuscated locations candidates, respectively. The K represents the mechanism that uses the probability $P(Z)$ to map an element in X to an element in Z. K is ϵ-geo-distinguishable, if and only if for all x, x' has:

$$d_{\mathcal{P}}\left(K(x), K\left(x'\right)\right) \leq \epsilon d\left(x, x'\right) \tag{1}$$

The $X_m = \{x_{m1}, \ldots, x_{mn}\}$ is the raw path that indicates the actual trajectories, while $Z_m = \{z_{m1}, \ldots, z_{mn}\}$ indicates the obfuscated trajectories. We utilize 2D Laplace noise $D_\epsilon(x)(z) = \frac{\epsilon^2}{2\pi} e^{-\epsilon d(x,z)}$ as obfuscation distribution in this paper. The reason is that the 2D Laplace noise ensures that z_{mi} is distributed around x_{mi}. The probability of z_{mi} with 2D Laplace noise decreases with the increasing of $d(x_{mi}, z_{mi})$ ($d(\cdot, \cdot)$ is the Euclidean distance in this paper). The ϵ-geo-distinguishable privacy condition is also satisfied with the 2D Laplace noise.

3.3 Adaptive Location Privacy-Preserving Scheme

A new adaptive location privacy-preserving scheme is proposed in this paper. The proposed scheme sets ϵ correlated to the generated obfuscated locations. By utilizing the proposed adaptive location privacy-preserving scheme, we boost the randomness of the noise generation, which increases the difficulty of inferring the actual location of a driver.

We configure ϵ into the high, medium, and low privacy levels. We set the average distance of the obfuscated location to r when there is a low level ϵ. A medium and a high level ϵ have an average distance of $1.5r$ and $2.25r$, respectively. The proposed scheme obfuscates every single location point in each continuous trajectory. As the starting location and destination are more sensitive to a driver, we set the highest privacy level for the two locations. For other locations x_i in the trajectory, the obfuscation parameters are related to the Euclidean distance $d(x_i, z_{i-1})$. We set two different thresholds $D1$ and $D2$ to divide $d(x_i, z_{i-1})$ into three types, where $D1 \neq D2$. When the value of $d(x_i, z_{i-1})$ is bigger than the values of $D1$ and $D2$, the correlation between x_i and z_{i-1} is weak. In this case, a low-level noise is added in the actual location data. When the value of $d(x_i, z_{i-1})$ is less than the values of $D1$ and $D2$, x_i and z_{i-1} is close in road networks (i.e., high correlation). Therefore, we add noise with a high privacy level when obfuscating actual locations in this scenario. Otherwise, we set ϵ as a medium privacy level in the proposed obfuscation scheme.

The proposed scheme reduces the correlation between x_i and z_{i-1}. Hence, the adversary cannot infer ϵ by analyzing the prior knowledge and the obtained obfuscated locations within a specific time window. The adversary cannot predict the driver's future locations because the value of ϵ is changing, which increases the difficulty of attacks.

The selection of ϵ in the proposed scheme aims to balance location privacy and data utility. A large amount of noise is required to achieve a high privacy level but leads to a low QoS of LBS. The different ϵ can provide customized geographic location accuracy which suit for various LBS requirements. For example, location-sensitive LBS (e.g., navigation) needs a high accuracy location data so that the scheme ought to utilize a high ϵ to provide a high data utility. For location-insensitive LBS, e.g., weather forecasts, the scheme can employ a low ϵ for a high privacy level.

The amount of the added noise is controlled in the proposed scheme to balance data utility and location privacy. we use z within a region based on x to set the upperbound of the QoS loss and that of the capability of location privacy, i.e., $d_{max}(x, z)$. If the distance $d(x, z)$ between the actual location x and the obfuscated location z exceeds $d_{max}(x, z)$, the proposed scheme will obfuscate the driver's actual location again.

The proposed scheme generates obfuscated locations $z_{m1}, \ldots, z_n$ and maps the obfuscated locations to the nearest road. Therefore, the proposed scheme obfuscates actual trajectories to reachable on-road locations.

3.4 Illegal Trajectories Detection Based on CNN

A two-dimensional convolutional neural network (2D-CNN) model is applied in the proposed scheme to detect illegal trajectories. The proposed model is shown in Fig. 1.

One-dimensional CNN (1D-CNN) consists of the convolutional layer, the subsampling layer, and the optional fully-connected layers. In convolutional neural networks, the convolutional layer is the major part that analyzes the input data to extract classification features. The Separate feature extractor contains multiple convolution kernels. The convolutional layer of the CNN model can extract the Spatial-temporal correlation of the trajectory. After extracting features in convolutional layer. Pooling step starts. The weight parameter redundancy is solved by the local connection and weight sharing. However, the over-fitting problems arise due to the CNN model degrades in the generalization performance. With the extracted features from the convolution layer and pooling, CNN model can reduce the data dimension while retaining the value of the principal feature map.

Our work improves the architecture of 1D-CNN with 2D-CNN model, which is widely utilized in classification of images (e.g., ResNet [24], VGG [25], and GoogleNet [26]). The proposed model is described as follows.

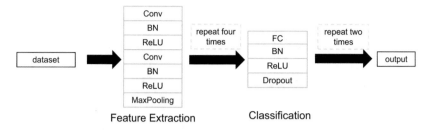

Fig. 1 Our CNN model architecture [23]

- In advance of the maximum pooling layer, two convolutional layers are employed in our model. Thus, the proposed model can extract features effectively.
- We add a normalization (BN) layer after the two convolutional layers to retain the spatial-temporal correlation of locations. The fully connected (FC) layers are combined with the dropout layers and BN in the proposed model to avoid the FC layers leading to over-fitting issue.

4   Evaluation

4.1   Original Dataset

We employ real-world road network information of Porto, which is extracted from OpenStreetMap (OSM) to evaluate the proposed scheme. OSM is popular in LBS applications, such as the route planning and the geocoding of address [27]. The extracted road network is shown in Fig. 2. The Portugal taxi trajectory dataset[2] whose recorded location data has 15 s time interval to build the trajectory, is used in our experiment.

4.2   Illegal Trajectories

As far as we know, no public dataset are labeled with illegal and legal trajectories. Three methods are popular to generate illegal trajectories according to legal trajectories.

---

[2] Portugal taxi trajectory dataset[Online]. Available: https://www.kaggle.com/c/pkdd-15-predict-taxi-service-trajectory-i.

Fig. 2 Generation of the road network. Upper: real road map in OSM; Bottom: the generated road network [23]

1. Insertion trajectory from other sources of trajectory dataset as the illegal trajectories [28].
2. Division the trajectories data into legal and illegal dataset [29].
3. Combination the above two methods [21].

In this paper, we generate illegal trajectories by utilizing legal trajectories. The generated illegal trajectories are employed in our classification experiments. Hence, legal and illegal trajectories in the training and detection come from the same dataset to reduce deviation.

4.3 Simulation Results

We evaluate the detection capability of the proposed scheme with the generated trajectory data and the public trajectory dataset of the Portugal taxi. In the experiment, we configure ϵ to control the level of Laplace noise.

4.3.1 Experimental Setting

We use the public real-world trajectory dataset and the synthetic data as follows.

Real-world trajectory: We employ the public trajectory dataset of Portugal taxi which has approximate 1.7 million trajectory data.

Synthetic data: We extract half of the trajectory data from the Portugal taxi dataset to generate the illegal data. In this paper, the illegal trajectories considered have two forms, speed anomaly and path anomaly. The two forms represent malicious actions, i.e., speeding and detour, respectively. The illegal trajectories are generated as follows:

- **Speed anomaly**: The maximum speeds of vehicles are limited in road networks. Malicious drivers drive faster than the limitations to obtain more profit within a period. The utilized trajectory dataset contains timestamp, we delete x_i in the selected continuous trajectory X with a certain probability and reassign the timestamp. The the trajectory X has a higher speed than the speed limitation.
- **Path anomaly**: Malicious drivers can also select a longer route than the recommended route, i.e., path anomaly. We utilize multiple legal trajectories (e.g., three trajectories X_a, X_b, X_c) to generate path anomaly illegal trajectories. The starting location and destination of X_a are denoted as x_{a1} and x_{an}, respectively, at shown in Fig. 3. We employ the trajectories X_b and X_c to intersect[3] the trajectory X_a at location x_{ai}, x_{aj}. We also combine the trajectories X_b and X_c at location x_{bl}. Then, we obtain an illegal trajectory which is a path anomaly. The starting location and destination of the generated trajectory are x_{a1} and x_{an}, respectively, but the route distance between x_{a1} and x_{an} is longer than it should be. In this paper, the length of the generated trajectories are set to be at least 1.6 times as long as that of the legal trajectories.

The above types of illegal trajectories include the most categories of malicious activities in the road networks. We use more than 600,000 legal and illegal trajectories to evaluate our proposed detection scheme.

4.3.2 Implementation

The configurations of the proposed mechanism that implemented in our experiments are as follows.

We use $\epsilon_0 = 0$ to stand for the non-protected situation. We configure the obfuscation radius as the average Euclidean distance between x_i and z_i when using noise level ϵ_i. When $\epsilon_0 = 0$, the average obfuscation radius is 0 m. We set ϵ_1 and ϵ_2 with obfuscation radii 100 m and 1000 m, respectively.

---

[3] The intersection stands for the points of the two trajectories whose distance are within a certain range.

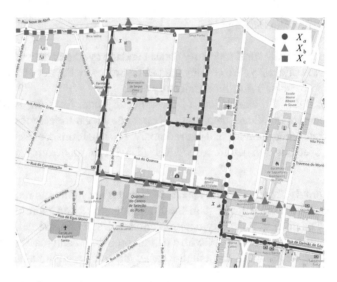

Fig. 3 Example of a anomaly generated path [23]

Table 1 Experimental accuracy

ϵ (m)	Average value of the noise radius	Accuracy (%)
ϵ_0	0	93.1
ϵ_1	100	86.1
ϵ_2	1,000	72.5

4.3.3 Experimental Results

We use Python to conduct the experiments. We take the average value of the experiment results after running the experiment for five times.

A large amount of noise decreases the accuracy rate of the CNN model, as shown in Table 1. We compare the receiver operating characteristics curve (ROC) with various setting of ϵ, as shown in Fig. 4. When applying ϵ_1 and ϵ_2, the proposed scheme achieves a high-value accuracy rate of 0.94 on the AUC score. Compared with that of ϵ_1, the AUC score of ϵ_2 has been reduced by 0.14–0.80.

4.4 Contrast

We use the same trajectory dataset as the existing related work [21]. When detecting illegal trajectories, the proposed scheme achieves higher accuracy rate than the scheme developed in [21]. The proposed scheme considers a more complex environment, i.e., road networks, than the 2D plane environment considered in [21]. The

Fig. 4 ROC under three different parameters [23]

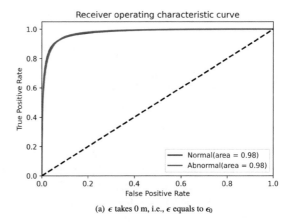

(a) ϵ takes 0 m, i.e., ϵ equals to ϵ_0

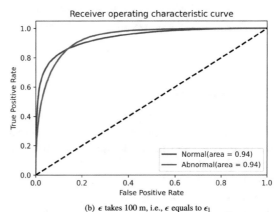

(b) ϵ takes 100 m, i.e., ϵ equals to ϵ_1

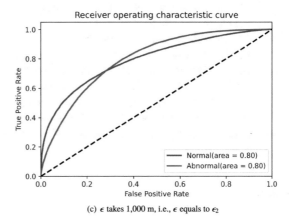

(c) ϵ takes 1,000 m, i.e., ϵ equals to ϵ_2

proposed scheme utilizes a different obfuscation process and recognition dataset, so we do not compare the accuracy of the two schemes. Compared with the schemes in [21], the proposed scheme has advantages as follows.

- The scheme in [21] employs fixed parameters in obfuscation process, while the proposed scheme dynamically calculate the parameters. The dynamically calculated parameters provide higher privacy protection capability and high data utility than the scheme in [21].
- The scheme in [21] manually inject trajectories to generate illegal trajectories. The illegal trajectories trajectories generated in this paper are closer to the real world than that of the scheme in [21]. The generated illegal trajectories in this paper are indistinguishable from the real trajectories which increases the difficulty to detect illegal trajectories. Under the strict assumption, the proposed scheme still achieve a higher accuracy rater than the existing work [21].
- We employ the real-world dataset and road networks to evaluate the proposed scheme. Thus, the proposed scheme in this paper has practical meaning.

5   Conclusions

In this paper, we first propose a new scheme to adaptively protect location data in real-world road networks, which is an important scenario of smart cities. Then, we proposed an illegal trajectory detection scheme to detect illegal locations in the case that all drivers are protected in road networks. The privacy parameters of the proposed scheme was calculated by considering the correlation of the actual location and the obfuscated location. Thus, the adversary cannot infer utilized privacy parameters and the actual locations. We generated illegal trajectory data with speed anomaly and path anomaly to simulate the real-wold malicious driving. The 1D-CNN model with 2D-CNN architecture is proposed in detecting illegal trajectories. According to our experiment results, the proposed detection scheme achieve better performance (e.g., the AUC score is above 0.94) than the existing works in road networks.

In the future work, we will balance the data utility and location privacy to maximize the data availability while satisfying the requirements of drivers' privacy. Moreover, we will assess the privacy levels of driver's privacy and develop a new system to protect location data privacy with the capability to handle most drivers' requirements.

References

1. Teng H, Dong M, Liu Y, Tian W, Liu X (2021) A low-cost physical location discovery scheme for large-scale internet of things in smart city through joint use of vehicles and uavs. Futur Gener Comput Syst 118:310–326

2. Sharma L, Javali A, Nyamangoudar R, Priya R, Mishra P, Routray SK (2017) An update on location based services: Current state and future prospects. In: 2017 international conference on computing methodologies and communication (ICCMC). IEEE, pp 220–224
3. Wang Y, Cai Z, Tong X, Gao Y, Yin G (2018) Truthful incentive mechanism with location privacy-preserving for mobile crowdsourcing systems. Comput Netw 135:32–43
4. Liu Z, Lei W, Ke J, Wenlei Q, Wang W, Wang H (2019) Accountable outsourcing location-based services with privacy preservation. IEEE Access 7:117258–117273
5. Xiong Z, Cai Z, Han Q, Alrawais A, Li W (2020) Adgan: protect your location privacy in camera data of auto-driving vehicles. IEEE Trans Industr Inf 17(9):6200–6210
6. Andrés ME, Bordenabe NE, Chatzikokolakis K, Palamidessi C (2013) Geo-indistinguishability: Differential privacy for location-based systems. In: Proceedings of the 2013 ACM SIGSAC conference on computer & communications security, pp 901–914
7. Al-Dhubhani R, Cazalas JM (2018) An adaptive geo-indistinguishability mechanism for continuous LBS queries. Wirel Netw 24(8):3221–3239
8. Adegoke Y (2017) Uber drivers in lagos are using a fake GPS app to inflate rider fares. Quartz Africa, November, 13
9. Ge Y, Xiong H, Liu C, Zhou Z-H (2011) A taxi driving fraud detection system. In: 2011 IEEE 11th international conference on data mining. IEEE, pp 181–190
10. Targio Hashem IA, Chang V, Badrul Anuar N, Adewole K, Yaqoob I, Gani A, Ahmed E, Chiroma H (2016) The role of big data in smart city. Int J Inf Manag 36(5):748–758
11. Wang S, Qin H, Sun Y, Huang J (2018) Privacy preservation in location-based services. IEEE Commun Mag 56(3):134–140
12. Dwork C (2006) Differential privacy, vol 2006, pp 1–12. ICALP
13. Yu L, Liu L, Pu C (2017) Dynamic differential; location privacy with personalized error bounds. In: Network and distributed system security symposium (NDSS)
14. Shokri R, Theodorakopoulos G, Le Boudec J-Y, Hubaux J-P (2011) Quantifying location privacy. In: 2011 IEEE symposium on security and privacy. IEEE, pp 247–262
15. Xiao Y, Xiong L, Zhang S, Cao Y (2017) Loclok: location cloaking with differential privacy via hidden Markov model. Proc VLDB Endowment 10(12):1901–1904
16. Chen C, Zhang D, Castro PS, Li N, Sun L, Li S (2011) Real-time detection of anomalous taxi trajectories from gps traces. In: International conference on mobile and ubiquitous systems: computing, networking, and services. Springer, pp 63–74
17. Mao J, Wang T, Jin C, Zhou A (2017) Feature grouping-based outlier detection upon streaming trajectories. IEEE Trans Knowl Data Eng 29(12):2696–2709
18. Yao D, Zhang C, Zhu Z, Qin H, Wang Z, Huang J, Bi J (2018) Learning deep representation for trajectory clustering. Expert Syst 35(2):e12252
19. LeCun Y, Bottou L, Bengio Y, Haffner P (1998) Gradient-based learning applied to document recognition. Proc IEEE 86(11):2278–2324
20. Nikhil N, Morris BT (2018) Convolutional neural network for trajectory prediction. In: Proceedings of the European conference on computer vision (ECCV) workshops
21. Suo D, Elena Renda M, Zhao J (2021) Quantifying the tradeoff between cybersecurity and location privacy. arXiv:2105.01262
22. Zurbarán M, Avila K, Wightman P, Fernandez M (2015) Near-rand: noise-based location obfuscation based on random neighboring points. IEEE Latin Am Trans 13(11):3661–3667
23. Zhao Y, Ma B, Wang Z, Liu Z, Zeng Y, Ma J (2022) Trajectory obfuscation and detection in internet-of-vehicles. In: 2022 IEEE 25th international conference on computer supported cooperative work in design (CSCWD). IEEE, pp 769–774
24. He K, Zhang X, Ren S, Sun J (2016) Deep residual learning for image recognition. In: Proceedings of the IEEE conference on computer vision and pattern recognition (CVPR), pp 770–778
25. Simonyan K, Zisserman A (2014) Very deep convolutional networks for large-scale image recognition. arXiv:1409.1556
26. Selvaraju RR, Cogswell M, Das A, Vedantam R, Parikh D, Batra D (2017) Grad-cam: visual explanations from deep networks via gradient-based localization. In: Proceedings of the IEEE international conference on computer vision (ICCV), pp 618–626

27. Maier G (2014) Openstreetmap, the wikipedia map. Region 1(1):R3–R10 Dec
28. Oh M-H, Iyengar G (2019) Sequential anomaly detection using inverse reinforcement learning. In: Proceedings of the 25th ACM SIGKDD international conference on knowledge discovery & data mining, pp 1480–1490
29. Gray K, Smolyak D, Badirli S, Mohler G (2018) Coupled IGMM-GANS for deep multimodal anomaly detection in human mobility data. arXiv:1809.02728

The Contribution of Deep Learning for Future Smart Cities

Hamidullah Nazari, Hasan Alkhader, A. F. M. Suaib Akhter, and Selman Hizal

Abstract With the advancement of science and technology, the idea of smart city is not a dream anymore; rather, the world is already enjoying its benefits. Computer and information technology have come up with Internet of Things (IoT), several data analysis technologies like big data analysis, deep learning, etc., taking smart city applications to a higher level. Current smart city applications offer various facilities to ensure more effective, time-saving services like traffic management, environmental and public safety and security, energy management, healthcare services, etc. The smart city's components firstly collect several data by using different IoT and other sensors. It requires processing those data to perform several actions to achieve the system's benefits. Deep learning methods that use artificial intelligence to make intelligent decisions are the most widely used data analysis technique used by the applications of smart cities. The data collected from the city through IoT or sensors are processed and analyzed, and then trainings are carried out with deep learning models. As a result of this training, effective solutions can be produced for different problems, and the workload of smart city members is facilitated. In this chapter, the categories and components of smart cities are presented, the necessity and applications of deep learning in smart cities are investigated, and the possible future enhancements that could be achievable by using deep learning for smart cities are enlightened.

H. Nazari · H. Alkhader
Department of Computer and Information Engineering, Sakarya University, Sakarya, Turkey

A. F. M. S. Akhter (✉) · S. Hizal
Department of Computer Engineering, Sakarya University of Applied Sciences, Serdivan, Sakarya 54187, Turkey
e-mail: suaibakhter@subu.edu.tr

S. Hizal
e-mail: selmanhizal@subu.edu.tr

© The Author(s), under exclusive license to Springer Nature Switzerland AG 2023
M. Ahmed and P. Haskell-Dowland (eds.), *Cybersecurity for Smart Cities*,
Advanced Sciences and Technologies for Security Applications,
https://doi.org/10.1007/978-3-031-24946-4_10

1 Smart City

Over 50% of the world's population currently lives in cities, and it is predicted that during the next 30 years, this percentage will rise to 68% [36]. The United Nations projects that by 2050, there will be 2.5 billion more people on Earth, and a massive number of them will enjoy the facilities of smart cities. Cities will face several issues due to this enormous increase, including how to manage and develop urban areas sustainably and ensure that residents have a high quality of life. In order to meet the many demands of this expanding population, the creation of smart cities should be seen as both an effective and urgent answer.

In smart cities, analyzing the growing variety of collected data by utilizing sensors from the real world is a complicated process. This data has been processed and analyzed using advanced technologies with various deep learning methods. Thus, applications developed using deep learning methods in smart cities help societies live at a reliable and prosperous level. Because smart cities are complex systems, it is almost impossible to explain all their applications. This study has tried to give information under several categories about applications developed with deep learning methods in smart cities.

Advances in information and communication technology (ICT), data analytics, and machine learning have made smart cities a reality. In a smart city, many IoT sensors are deployed across many locations to collect data [7]. Where this data collected by sensors is analyzed to determine the strength of the technological infrastructure in cities, these technologies are used in several areas, including environmental issues, smart public transportation, to protect the safety of homes from burglaries and theft, and to provide the safety of citizens as well. Deep learning is one of the essential machine learning techniques, as it has an influential role in understanding and analyzing data and using many algorithms for classification and prediction.

1.1 Categories of Smart City

Based on a variety of research and analysis, four categories of smart cities were revealed, according to the international research team [34]. These four categories may help in proper planning by comparing these types with the conditions and infrastructure of cities and starting work in proportion to the social and economic situation (presented in Fig. 1); these are:

- **Essential Services:** This category falls within the services and basic needs of cities and people through mobile phone networks within the emergency health system and digital health care services; Examples include Tokyo and Copenhagen.
- **Smart transportation:** This model may be suitable for those densely populated cities that suffer from traffic congestion and a huge number of cars, as cities in this group are working to solve these problems through smart public transportation and

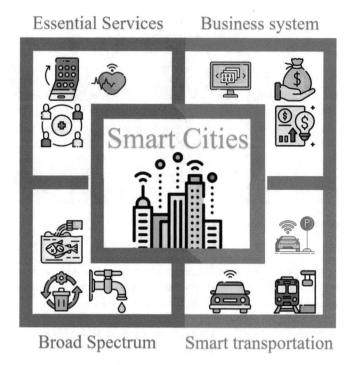

Fig. 1 Categories of Smart Cities

the development of smart self-driving cars, in addition to smart parking systems based on information and communication technologies, for example, "Dubai".

- **Broad Spectrum:** These cities fall within the broad-spectrum model that focuses on urban services, such as water, sanitation, and waste management systems, and also tries to solve pollution problems and place great importance on civic participation, such as Barcelona.
- **Business system:** This model stimulates economic activity based on the capabilities of technology and information and trains people on digital programs and skills to help build a highly trained generation of technology to develop and promote businesses such as Amsterdam.

1.2 Components of Smart City

Smart cities are not limited to sensors to automate daily life. It's much more than that. By investigating different researches, it has been found that there are four primary components of smart cities, without which an integrated smart city cannot be built Fig. 2.

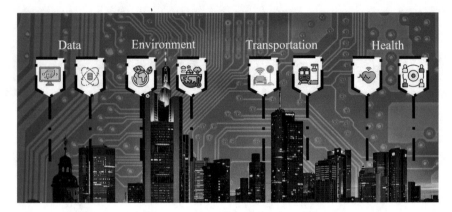

Fig. 2 Components of Smart Cities

1. **Analysis with collected data:** Modern technologies and sensors in smart cities collect a considerable amount of data daily. Still, the most important thing is knowing how to use, analyze and benefit from this data to improve the quality of life, as the considerable amount of data means nothing if it is not used correctly.
2. **Transportation Solutions:** In smart cities, there is an interconnected network between public transportation, infrastructure, and citizens to improve and develop the mobility system. As the use of smart devices and sensors in movement and transportation can play a significant role in:

   • Passenger and pedestrian safety
   • Suggest the best time to travel
   • Eliminate congestion and waste of time for transportation and passengers
   • Offer alternative options.

"We are able to understand the landscape of a community and can bring some technologies or work with partners who can offer technologies that can help solve these transportation issues" said Dominik Garcia, Senior Market Manager in Battelle's Transportation business.

3. **Health/Smart healthy cities:** There are many challenges in the field of health care, the most important of which is the provision of health care, especially for remote areas. As one of these challenges is the use of sensitive devices that people wear daily to detect some diseases such as diabetes. The biggest challenge is solutions for pregnant women who live in disadvantaged communities and facilitating the process of conducting periodic prenatal checkups. There is also a challenge linking the environment and health, as using devices to monitor air quality and warn people if there are any allergens or suffocation in the air helps people to plan their days and holidays accurately.

4. **Environment:** Solutions in smart cities help to cope with changing climate and environmental conditions. Since the focus is on preparedness, avoiding hurricanes or floods is impossible, but the extent of readiness dramatically reduces the damage. The sensors in the water play a role in determining the locations of the flood plains or the possibility of their occurrence. Also, by taking advantage of previous data related to environmental conditions, sensors inside forests can predict potential fires.

2 Deep Learning

Deep Learning is a sub-field of machine learning that involves algorithms boosted by the function and structure of the brain called "deep neural networks." Deep learning is beneficial when the data is complicated and massive data sets are available. With the development of deep learning, a new era has opened to the world. Large companies have started to use deep learning in various fields. The bulk of internet firms and high-end consumer products now employes employers who are qualified to participate in work with deep learning algorithms. For example, Amazon engineers developed computer vision algorithms that use deep learning to accurately utilize, Go stores to associate transactions with the right customer. To analyze text using deep learning in online conversion by Facebook. Companies such as Google, Baidu, and Microsoft utilized deep learning models for image search [6] and machine translation [38].

2.1 Neural Networks

The idea of deep learning has emerged with the invention of neural networks. Neural networks are inspired by the human brain and aim to create a model that performs functions like our brain. The purpose of neural networks is to imitate the human brain and create a system that can process information like the human brain. It can be said that the structure that forms the skeleton of deep learning is neural networks. So, deep learning is diversified according to the type of neural network. Here, we tried to explain briefly and focused on three basic types of neural networks.

2.1.1 Deep Neural Network (DNN)

DNN architecture and name were inspired by the human brain and emerged due to the mathematical modeling of the learning process. Imitate how biological neurons communicate with each other. In other words, the digital modeling of biological neuron cells and the synaptic bond these cells establish. Network layers of neurons connected to each other are called nodes. These nodes as Fig. 3 primarily consist of input, one or more hidden, and output layers.

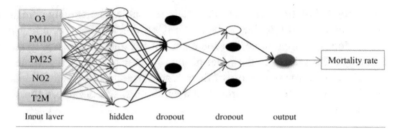

Fig. 3 Overview of Deep Neural Network architecture [24]

DNNs are used in every part of life to solve problems, especially in healthcare, automotive, electronics, space sciences, banking, finance, military, and so on.

2.1.2 Convolution Neuron Network (CNN)

The development of deep learning has led to the rapid growth of various fields such as computer vision, image analysis, etc. All these developments have been made possible by the emergence and progress of convolutional neural networks. The convolutional neural network is a deep learning model or algorithm that can recognize and classify features in images. It is a multi-layered neural network designed to perform many operations such as image and text analysis, classification, segmentation, and object recognition. CNN generally consists of many convolutional layers, relu layer, pooling layer, fully connected layer, and activation functions [28]. The architecture of CNN has been shown in Fig. 4.

To summarize CNN, the convolution, Relu, and pooling layers extract the feature maps of the input images and are called feature extraction layers. The flatten layer converts the outputs of the previous layers to vectors. The fully connected layer determines the probability value of which class as output, using the Dense and Activation functions of the features extracted from the image.

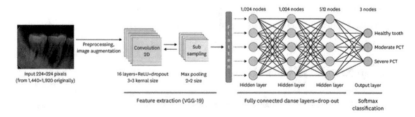

Fig. 4 Overview of Convolution Neuron Network architecture [21]

2.1.3 Recurrent Neural Networks (RNN)

RNNs use the previous output to predict the next output for sequential data. Recurrent neural networks consist of loops. Loops that store information about words from a while ago are stored in hidden neurons for predicting outputs. The output made for t time information from the hidden layer is being sent to the hidden layer back [11]. A recurrent neuron's output is sent to the next layer after all time information is provided. This output consists more of general information compared to the previous output. Previous information is stored for a longer time. The weights of the unexposed network need to be updated for the error to backpropagate. This is called backpropagation in time, the architecture shown in Fig. 5.

RNNs are used in different fields such as prediction problems, language modeling and generating text, machine translation, speech recognition, image descriptions, video tagging, text summarization, etc. Other deep learning models given in Fig. 6 are widely used in various fields, especially to solve the multiple problems of smart cities.

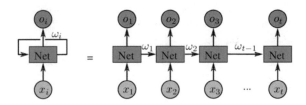

Fig. 5 Overview of Recurrent Neural Networks architecture [33]

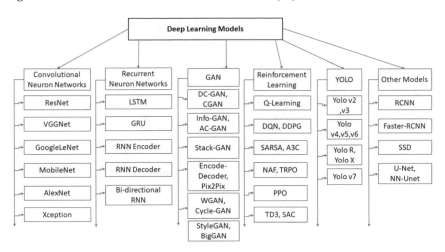

Fig. 6 List of the Models of Deep learning methods

2.2   The Contribution of Deep Learning in Smart City

The enhancement of smart city-related applications is greatly increased by combining with extraordinary deep learning services. Figure 7 represents the frequency of utilizing deep learning in smart cities.

A short overview of the contributions is listed with some examples in this section.

2.2.1   Intelligent Mobility and Transportation

The transportation system is anchored through artificial intelligence, cloud platforms, and connecting people, vehicles, and infrastructure.

There are many traffic-related problems. Where researchers have used deep learning, data analysis, and communication technologies to connect people, methods, and vehicles to solve various issues, this work aims to create a safe environment around vehicles [3]. The authors devised a support system for safe driving to prevent drivers from falling asleep or falling asleep [39]. In another research, a pedestrian detection system was worked on through a 3D stereo camera that can detect pedestrians in the area using artificial intelligence. This system helps drivers who drive in congested, unseen areas and blind spots [32].

2.2.2   Infrastructure

Research confirms that by 2030 the proportion of people living in cities will exceed 60%. So this will lead to many problems in cities, the most important of which are overcrowding and sustainability. Addressing these challenges is very important

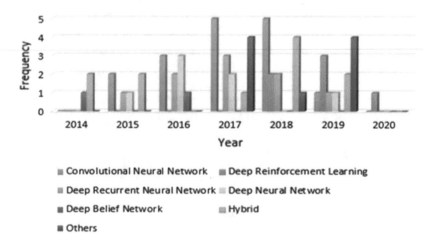

Fig. 7 Frequency of Deep learning methods in Smart City applications [25]

because of the rapid population growth. Where the strength of the infrastructure of cities is the first point of the solution, and it plays an essential role in facing these challenges.

The authors incorporate deep learning techniques into an intelligent infrastructure model. This system monitors traffic rate and power consumption and then makes a decision based on severity [30]. In another study, the researchers suggested a mechanism to deal with the huge amounts of data generated by the sensors using deep learning [40].

2.2.3   Smart Health Centers

Artificial intelligence has many smart and advanced health solutions, especially the modern concept of deep learning. Cancer is one of the most challenging diseases that humanity suffers. Starting treatment after the disease is not always guaranteed, so using deep learning and transfer learning models has been very useful in classifying breast cancer. Through data and images from breast cytology, the classification and prediction resulting from the breast images were more accurate than traditional deep learning structures [18]. On the other hand, deep learning models are also used to analyze the sounds of newborns through audio simulation of the crying sound to know and predict the newborn's health status and take the necessary measures in the event that any health abnormality is detected [22]. The use of deep learning methods encouraged the development of systems to predict possible future diseases in humans through electroencephalography based on deep learning technology, which can pick up brain signals to detect the possibility if a person suffers from diseases [2].

2.2.4   Smart Technologies to Develop the Education System

Distance learning has become a necessary need due to the development of technology, especially after the Corona pandemic, which forced students to complete their education remotely, as one of the techniques that were used is the method of Online Flipped Classroom Learning Method (OFCLM). This method has proven to increase students' thinking skills through brainstorming lectures via the Internet [31]. On the other hand, the researchers applied a system using computer vision called the face detection system by using features and emotions on the students' faces, whether the student is happy, sad, sleepy or focused. This vast data contributed to the analysis and improvement of the education system to suit students [12].

2.2.5 To Improve Government Policies for Urban and Smart Management

People are moving around cities more is one of the main reasons for building more intelligent and smarter cities. The researchers review the management concept, which helps analyze government policies and the characteristics of smart cities [10]. Public opinion is an essential factor in improving government policies toward urbanization.

3 Application of Deep Learning in Smart City

The smart city requires using advanced and scalable interconnected technologies to better solve our age's problems. With the help of these new technologies, it makes it possible to analyze in real-time various large and complex data quickly and robustly through deep learning applications. Thanks to deep learning applications, it provides convenience in many areas, such as distributing energy to communities more efficiently, analyzing environmental factors, reducing traffic congestion, and improving air quality. Deep learning applications developed for smart cities are examined in several categories, such as security and safety, pandemic control strategies, traffic monitoring tasks, etc.

3.1 *Smart City Applications for Security and Safety*

- **Environment monitoring and Person detection:** Human recognition is widely used in smart cities. In these applications, human activity detection is interpreted with real-time video analytics in areas such as train stations and airports. Deep learning applications are used to locate the target and automatically detect intrusion events on ring roads in real-time. Thus, environmental monitoring systems provide a wide safe area [27].
- **Recognizing violent and perilous circumstances:** With smart city applications, it aims to ensure the safety of the public and protect them from dangerous situations such as fights, thefts, and violence against women by using smart video surveillance [5]. Deep learning methods allow it to automatically follow and recognize human actions and detect many dangerous events.
- **Real-world Anomaly Detection in Surveillance Videos Using Deep Reinforcement Learning:** In order to increase security in public places, using the sensors of security cameras, video analysis with reinforced learning was carried out to ensure human safety in areas such as banks, intersections, shops, streets, parking lots, etc. [1].
- **Safety and adherence Control and Inspection Using Deep Learning:** In smart cities, a wide variety of compliance monitoring situations of deep learning-based applications automatically control the area, just like job security specialists. It

monitors compliance by detecting the conditions of the employees, including helmets, safety glasses, vests, hand gloves, steel-toed boots, etc., during work in the factory or construction areas [14].

- **Prevention of Suicide in Public Areas:** Deep learning-based applications have been developed to automatically prevent suicides in public spaces by detecting human movements and behaviors by extracting image features using advanced cameras. Deep learning applications have made it possible for it helps to identify people who attempt to harm themselves and initiate an intervention. Its primary purpose is the development of automated detection systems for early intervention, and pre-experimental behavior detection [17].

- **Crowd Catastrophe Prevention Software:** Crowd disaster prevention systems are used to improve public safety. Such applications focus on crowd scene and behavior analysis using deep learning models. For example, it is used to count people and predict a crowded environment in real-time with multiple cameras. Therefore, vision-based crowd disaster prevention systems are used to increase public safety [4].

- **Deep Learning Based Weapon Identification and Reporting:** Controlling crime rates in densely populated cities is a complicated process. Surveillance processes can be performed in real-time in public places using deep learning methods. These models automatically analyze the video stream to detect objects appearing in the camera feed. Thus, it detects the armed person, sends a notification to the central system, and helps the person to be approached quickly by following the suspect with deep sort, and tracking algorithms [26].

3.2 Traffic Monitoring Using Deep Learning in Smart City

- **Traffic Monitoring:** Traffic conditions can be analyzed and predicted using advanced deep learning models. Traffic data usually consists of the number, frequency, and direction of cars collected from security cameras. Deep learning networks are helpful for vehicle counting and can detect different vehicle types in heavy traffic situations. Thus, monitoring traffic waiting times and controlling traffic flows can be easily observed [23].

- **Driver Assistance Systems:** Many systems have been proposed in Smart City Systems that help the driver and vehicle recognize dangerous traffic situations. These systems generally aim to control safety and comfort. In vehicle systems, traffic sign detection and driver behavior monitoring are used to warn drivers in situations such as fatigue and sleepiness [35].

- **Detecting the Violations of Traffic Laws:** Visual data analysis has become possible using modern deep learning methods to detect violations of traffic rules in smart cities. For example, the detection of motorcycles and cyclists without helmets, the detection of situations that lead to traffic accidents, and the detection of vehicles standing in dangerous places are carried out [20].

- **Implementation of Roadside Surveillance:** Automatic surveillance of the occupation of roadsides can be realized with deep learning methods. Using these methods, systems that perform real-time monitoring and detection of unloading and loading operations of vehicles such as trucks have been developed [13].
- **Detecting Parking Lot Occupancy:** Today, real-time occupancy monitoring in parking lots has become significant with the increase in vehicles. Deep learning-based systems that help them find empty parking spaces through multiple cameras have been developed. These systems allow vacant parking lots to be easily found and tracked [16].
- **Control of Drivers with Drone:** Using seat belts, one of the most basic rules of a safe journey, saves the lives of the driver and passengers even in the most severe accidents. The images taken from the drone can be controlled remotely to automatically detect, with deep learning methods, whether the drivers are wearing their seat belts [37].

3.3   Measures for a Smart City to Prevent the COVID-19 Pandemic

- **Controlling Social distancing in public places:** Social distancing is crucial for communities to prevent the spread of pandemic situations such as COVID-19. It can reduce virus transmission within a community by maintaining social distance in places such as shopping malls, religious places of worship, schools, and public transportation. Deep learning techniques have been utilized to limit social distance and stop the pandemic's spread [15].
- **Automatic Mask Detection:** One factor that reduces the transmission of the pandemic is that people wear masks. Wearing a face mask and following social distancing rules in public can minimize the spread of the virus. It can facilitate mass surveillance by using deep learning algorithms to monitor conditions automatically and send alerts in case of non-compliance with quarantine measures [29].

3.4   Measurements of Air Quality

- **Analyzing the Quality of Air Pollution:** Air pollution is the presence of a higher than average amount and density of foreign substances in the air that adversely affect the health of living things. Air pollution quality prediction has been carried out with deep learning models using IoT data in smart cities [19]. As a result of the prediction, municipalities conduct various studies to minimize the city's air pollution.

- **Ozone Concentration Prediction in a Smart City:** One of the air pollutants having the most negative effects on human health is ozone. For ozone level estimation, predictions of pollution and weather correlations are realized by using deep learning models. Like these studies are vital in smart cities, where innovative pollution systems predict air pollution levels [9].

4 Future Opportunities

Despite the tremendous development and the role that deep learning plays in developing and facilitating life in smart cities, many challenges still exist in developing applications for smart cities. The most important of these challenges is to find and analyze huge amounts of data to use in sensitive applications, as in some cases, the amount of data is insufficient. And the development is still going on, as there are quite a few applications we can see in smart cities during the coming period, especially in the health field, where devices for detecting chronic and dangerous diseases. And also improving the quality of education in schools and universities, as researchers are working on finding algorithms to develop and enhance the distance learning process, especially after the Corona crisis. Rehabilitation of the infrastructure for smart self-driving projects for vehicles, as the idea of self-driving still needs to be improved from the technical side on the one hand and from the infrastructure side, such as highways on the other. It is possible to summarize the future research opportunities as followings:

- According to research performed by Chiroma et al. in [8], in many of the applications of smart cities, it has been found that nature-inspired deep learning algorithms perform well than typical methods. It requires more experiments and finding better ways to produce better outcomes.
- Unclassified data remains a problem for deep learning methods, and in the case of smart city applications, this problem is very obvious. Inside a system are mixtures of components, each producing different types of information and in various formations. For example, a vehicular application system has mixtures of autonomous cars with semi-autonomous and non-autonomous vehicles. So, there are research opportunities to handle this type of problem.
- The dynamic nature of technology is another challenge for smart city-related researchers and developers. Continuous updating technologies can make some existing applications obsolete with time. Thus, standardization and compatibility management will remain a challenge for the researchers.
- The scope and opportunities of deep learning methods are enormous. Still, many DL algorithms are not implemented or tested for smart city applications. It may be possible to find out more efficiency while utilizing those methods.
- Deep learning algorithms are updating day by day. So, using updated techniques in smart city applications could be more effective than the currently used methods in smart city applications.

- Real-world implementation and experimental results are still unavailable in deep learning (used for smart cities). So, more experiments, results, and performance analyses remain as potential future works.

5 Conclusion

One of the cornerstones of smart cities is applications developed with deep learning methods. Big data obtained with IoT devices are trained in deep learning methods using advanced cutting-edge technologies. Intelligent systems are developed using these methods and play an important role to facilitate different components of smart cities. This chapter has made an effort to provide information on applications that are being developed under various categories such as security and safety, traffic monitoring, prevention of the COVID-19 pandemic, and air quality for smart cities and that are anticipated to be done in the future. Our research aims to inspire further smart city researchers to investigate this fascinating area of study further and produce more inventive and technically useful deep learning models particularly created for smart city applications.

References

1. Aberkane S, Elarbi M (2019) Deep reinforcement learning for real-world anomaly detection in surveillance videos. In: 2019 6th international conference on image and signal processing and their applications (ISPA). IEEE, pp 1–5
2. Alhussein M, Muhammad G, Hossain MS (2019) Eeg pathology detection based on deep learning. IEEE Access 7:27781–27788
3. An C, Wu C (2020) Traffic big data assisted v2x communications toward smart transportation. Wirel Netw 26(3):1601–1610
4. Atitallah SB, Driss M, Boulila W, Ghézala HB (2020) Leveraging deep learning and IoT big data analytics to support the smart cities development: review and future directions. Comput Sci Rev 38:100303
5. Baba M, Gui V, Cernazanu C, Pescaru D (2019) A sensor network approach for violence detection in smart cities using deep learning. Sensors 19(7):1676
6. Baker GA, Wang J, Fan M, Weatherley LR (2009) Foreword
7. Bhattacharya S, Somayaji SRK, Gadekallu TR, Alazab M, Maddikunta PKR (2022) A review on deep learning for future smart cities. Internet Technol Lett 5(1):e187
8. Chiroma H, Gital AY, Rana N, Abdulhamid SM, Muhammad AN, Umar AY, Abubakar AI (2019) Nature inspired meta-heuristic algorithms for deep learning: recent progress and novel perspective. In: Science and information conference. Springer, pp 59–70
9. Ghoneim OA, Manjunatha B et al (2017) Forecasting of ozone concentration in smart city using deep learning. In: 2017 International Conference on Advances in Computing. Communications and Informatics (ICACCI). IEEE, pp 1320–1326
10. Grossi G, Meijer A, Sargiacomo M (2020) A public management perspective on smart cities:'urban auditing' for management, governance and accountability
11. Gul M, Celik E, Mete S, Serin F (2020) Computational intelligence and soft computing applications in healthcare management science. IGI Global

12. Gupta SK, Ashwin T, Guddeti RMR (2019) Students' affective content analysis in smart class-room environment using deep learning techniques. Multimedia Tools Appl 78(18):25321–25348
13. Ho GTS, Tsang YP, Wu CH, Wong WH, Choy KL (2019) A computer vision-based roadside occupation surveillance system for intelligent transport in smart cities. Sensors 19(8):1796
14. Jindal A, Aujla GS, Kumar N, Prodan R, Obaidat MS (2018) Drums: demand response man-agement in a smart city using deep learning and svr. In: 2018 IEEE global communications conference (GLOBECOM). IEEE, pp 1–6
15. Kalra J, Pant R, Gupta S, Kumar V (2021) Social distance monitoring in smart cities using IoT. In: Green internet of things for smart cities. CRC Press, pp 135–145
16. Karakaya M, Akıncı FC (2018) Parking space occupancy detection using deep learning meth-ods. In: 2018 26th signal processing and communications applications conference (SIU). IEEE, pp 1–4
17. Karthick A, Sakthi T (2021) Standalone PV-wind-DG-battery hybrid energy system for zero energy buildings in smart city Coimbatore, India. Springer International Publishing, pp "55–63"
18. Khan S, Islam N, Jan Z, Din IU, Rodrigues JJC (2019) A novel deep learning based framework for the detection and classification of breast cancer using transfer learning. Pattern Recogn Lett 125:1–6
19. Kök I, Şimşek MU, Özdemir S (2017) A deep learning model for air quality prediction in smart cities. In: 2017 IEEE international conference on big data (Big Data). IEEE, pp 1983–1990
20. Kumar A, Kundu S, Kumar S, Tiwari UK, Kalra J (2019) S-tvds: Smart traffic violation detection system for Indian traffic scenario. Int. J. Innov. Technol. Explor. Eng. (IJITEE) 8(4S3):6–10
21. Lee JH, Kim D, Jeong SN, Choi SH (2018) Diagnosis and prediction of periodontally com-promised teeth using a deep learning-based convolutional neural network algorithm. J Period Implant Sci 48(2):114–123
22. Liang P, Sun G, Wei S (2019) Application of deep learning algorithm in cervical cancer MRI image segmentation based on wireless sensor. J Med Syst 43(6):1–7
23. Lingani GM, Rawat DB, Garuba M (2019) Smart traffic management system using deep learn-ing for smart city applications. In: 2019 IEEE 9th annual computing and communication workshop and conference (CCWC). IEEE, pp 0101–0106
24. Maharani D, Murfi H (2019) Deep neural network for structured data-a case study of mortality rate prediction caused by air quality. In: Journal of physics: conference series. IOP Publishing, vol 1192, p 012010
25. Muhammad AN, Aseere AM, Chiroma H, Shah H, Gital AY, Hashem IAT (2021) Deep learning application in smart cities: recent development, taxonomy, challenges and research prospects. Neural Comput Appl 33(7):2973–3009
26. Muheidat F, Tawalbeh L (2021) Artificial intelligence and blockchain for cybersecurity applica-tions. In: Artificial intelligence and blockchain for future cybersecurity applications. Springer, pp 3–29
27. Nayak R, Behera MM, Pati UC, Das SK (2019) Video-based real-time intrusion detection system using deep-learning for smart city applications. In: 2019 IEEE international conference on advanced networks and telecommunications systems (ANTS). IEEE, pp 1–6
28. Nazarl H, AkgÜn D (2020) A deep learning model for image retargeting level detection. In: 2020 4th international symposium on multidisciplinary studies and innovative technologies (ISMSIT), IEEE, pp 1–4
29. Rahman MM, Manik MMH, Islam MM, Mahmud S, Kim JH (2020) An automated system to limit covid-19 using facial mask detection in smart city network. In: 2020 IEEE international IOT, electronics and mechatronics conference (IEMTRONICS). IEEE, pp 1–5
30. Serrano W (2019) Deep reinforcement learning algorithms in intelligent infrastructure. Infras-tructures 4(3):52
31. Shu F, Zhao C, Wang Q, Huang Y, Li H, Wu D (2019) Distance learners' learning experience and perceptions on the design and implementation of an online flipped classroom learning model. In: 2019 eighth international conference on educational innovation through technology (EITT). IEEE, pp 7–11

32. Solmaz G, Berz EL, Dolatabadi MF, Aytaç S, Fürst J, Cheng B, Ouden Jd (2019) Learn from iot: pedestrian detection and intention prediction for autonomous driving. In: Proceedings of the 1st ACM workshop on emerging smart technologies and infrastructures for smart mobility and sustainability, pp 27–32
33. Tai L, Liu M (2016) Deep-learning in mobile robotics-from perception to control systems: a survey on why and why not. arXiv:1612.07139 1
34. Tang Z, Jayakar K, Feng X, Zhang H, Peng RX (2019) Identifying smart city archetypes from the bottom up: a content analysis of municipal plans. Telecommun Policy 43(10):101834
35. Tumen V, Yildirim O, Ergen B (2018) Recognition of road type and quality for advanced driver assistance systems with deep learning. Elektronika ir Elektrotechnika 24(6):67–74
36. UnitedNations (2018) 68% of the world population projected to live in urban areas by 2050, says un l un desa department of economic and social affairs. https://www.un.org/development/desa/en/news/population/2018-revision-of-world-urbanization-prospects.html. Accessed from 2022-08-04
37. Vattapparamban E, Güvenç I, Yurekli AI, Akkaya K, Uluağaç S (2016) Drones for smart cities: issues in cybersecurity, privacy, and public safety. In: 2016 international wireless communications and mobile computing conference (IWCMC). IEEE, pp 216–221
38. Yang S, Wang Y, Chu X (2020) A survey of deep learning techniques for neural machine translation. arXiv:2002.07526
39. Zhang R, Xie P, Wang C, Liu G, Wan S (2019) Classifying transportation mode and speed from trajectory data via deep multi-scale learning. Comput Netw 162:106861
40. Zhao L, Wang J, Liu J, Kato N (2019) Routing for crowd management in smart cities: a deep reinforcement learning perspective. IEEE Commun Mag 57(4):88–93

An EDGE Supported Ambulance Management System for Smart Cities

A. F. M. Suaib Akhter, Selman Hizal, Tawsif Zaman Arnob, Ekra Binta Noor, Ehsanuzzaman Surid, and Mohiuddin Ahmed

Abstract Current destruction because of COVID pandemic has force the researchers to think about the upgrading of the healthcare system. Involvement of updated technology can be one of the prospective field to be applied in the healthcare to enhance its efficiency. In this paper, an edge computing supported healthcare system is proposed where smart ambulances are utilized to perform primary diagnosis of the patients and send it to the medical center so that it is possible to organize necessary actions before the patient come. In case of patient coming from long distance it is possible to provide primary treatment for the patient by analyzing the diagnosis information sent from the carrier ambulance. Furthermore, the proposed EHealth system supports video call to establish a communication between the attendant of the ambulance and doctor in case of emergency. All the features together could provide life saving support for the patients especially to them whom are coming from rural area and it takes hours to reach to the hospital. A blockchain is utilized to preserve the sensitive data of the patient and edge computing is used to facilitate the complete process by providing computational supports. A proof-of-concept is presented to simulate the communication protocol through blockchain and performance of the EHealth system is presented in terms of time and storage consumption which shows that it require minimum time and storage to provide these productive support by the EHeath system.

M. Ahmed
Edith Cowan University, Joondalup, WA, Australia
e-mail: m.ahmed.au@ieee.org

A. F. M. S. Akhter (✉) · S. Hizal
Sakarya University of Applied Science, Sakarya, Turkey
e-mail: suaibakhter@subu.edu.tr; afmsuaibakhter@gmail.com

S. Hizal
e-mail: selmanhizal@subu.edu.tr

T. Z. Arnob · E. B. Noor · E. Surid
Islamic University of Technology, Gazipur, Bangladesh
e-mail: tawsifzaman@iut-dhaka.edu

E. B. Noor
e-mail: ekra@iut-dhaka.edu

1 Introduction

To enhance quicker access to medical history of patients, Electronic Health Records or EHRs, are already being integrated by hospitals in recent times. With the aid of Electronic Health Records, it is possible to better patient care, enhance the clinical performances, and promote healthcare research. Advancing to the future, healthcare driven data is certainly on a rise and to handle the data in an efficient manner, technologies such as a network with distributed data, parallel processing, building scalable storages, infrastructures, frameworks, etc. are required [1]. Moreover, the cost of infrastructure setup and maintenance are increasing as online based computing services are becoming popular because of their ease of use, security, scalability, reliability, interoperability, etc. Therefore, the healthcare sector can be benefited by using edge services without having the responsibility of maintenance and server setup. The problem of storing different types of data can be solved by using blockchain which is more secure, reliable and decentralized. But healthcare devices need a lot of computational power and scalable storage to perform complex operations. In this case, edge computing with cloud service can be used which will do computational work and provide secured space to store data. High speed 5G network can be used to transmit all sorts of data between the components (Blockchain, cloud and edge server) to reduce runtime and increase efficiency.

Edge computing is a service that provides distributed computational and storage services so that members can use them to perform complex computational tasks and store big data. By transferring tasks to the edge service providers, light weight devices like Internet of Things (IoTs), Internet of Vehicles (IoVs) can perform massive data computational tasks indirectly. By combining cloud with edge smart objects are now able to participate in complex operations without having powerful processing units or storage. In recent times, use of lightweight smart devices are increasing in the healthcare sector and most of them are using mobile edge computing to provide efficient and quick service [2]. 5G technology shows promise for the future to be integrated in intelligent healthcare systems, which can enhance spectrum efficiency as well as enable the upgraded high-speed communications.

For reliable, secured and efficient data storage and management the popularity of blockchain is increasing day-by-day. Blockchain is a distributed ledger which provides trustless security among its nodes. Although, it was first introduced to provide coin exchange support by removing the dependency on trusted third party, exceptional features like decentralization, flexibility, transparency, integrity, immutability, confidentiality, temper resistance, robustness, attack prevention capability, etc. make it popular in different sectors [3, 4]. Healthcare in one of the potential field which also take advantages of blockchain for different purpose like storing medical records, pharmaceutical supply chain, remote patient monitoring, health data analysis, etc. Fig. 1 illustrates the applications of blockchain in healthcare.

There are several areas of research in the healthcare sector and ambulance management system is one of the important part of this. But unfortunately, only a few number of research is found related to this field. In the current pandemic situation the neces-

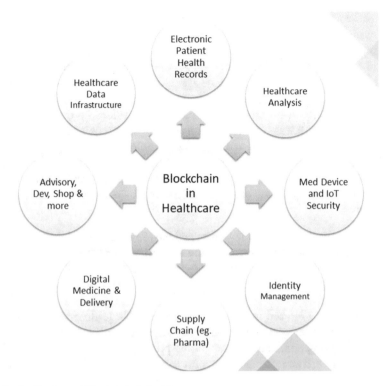

Fig. 1 Applications of blockchain in healthcare

sity of efficient ambulance management is coming forward. Previously ambulance was used only for the emergency situations but huge load of COVID-19 infected patients brings several problems in front. Now, patients can not get ambulance in emergency situation, if possible to manage one the hospital or nearby medical centers are full of patients and there is not space remaining for new one. The patients from rural areas suffer more in this situation. Thus, in this paper, we have proposed an efficient ambulance management system to minimize the struggles faced by patients in the COVID situation named EHealth. In the proposed method, a blockchain is implemented to store and manage information about the components of the system i.e., ambulance, medical centers and patients. A patient can ask for ambulance in case of an emergency or to test COVID. Blockchain manager will store and broadcast it to the nearby ambulance service providers. Interested ambulances will response to the request and the patient will select one of them for the service. After instant COVID test, if the result come as positive, the patient will send a request for medical center for admission to the blockchain and it will broadcast it to the medical centers of that area. Medical centers with patient acceptance capability will response and patient can select if there are multiple options. In case of emergency and if the medical center is far from the patients place, it may require live physician support which can be provided by the proposed system with the help of 5G high speed internet. Additionally, ambulances with primary diagnostic facilities can perform the tests on the way to the hospital and send data to the medical center which will be managed by the blockchain. By this way, the proposed system can manage ambulance, medical centers and patients together to provide faster and efficient facility in the critical situations.

Online ambulance management is comparatively new idea for the researchers as ambulance service is always considered as a service which requires instant response. But, because of the COVID-19 and related pandemic situation ambulance service providers are unable to ensure instant service. Patients from rural areas are suffering more because of this. Thus in this paper, we propose a blockchain based ambulance management system by adding edge and cloud service. Contribution of the paper are the followings:

- A blockchain based system which stores patients, ambulances, medical centers and doctors information to establish a communication between them. Service providers (ambulances, medical centers, doctors) can broadcast the details like facilities, availability, expenses, etc. and patients can get necessary supports by using this platform.
- Smart ambulance is used in this system that are able to do primary diagnosis, have high speed internet connectivity and required hardware to provide video conferencing supports. Patients can get the primary test reports before reaching to the hospital and can get online support from a doctor while on the way to the hospital.
- All the communication between service providers and patients will be stored in the blockchain to get the security features of it and to perform complex calculation and store all the related information edge computing will be used together with cloud. Participants of the system (service providers and patients) will use high speed 5G internet to send and receive information from the infrastructures (blockchain, edge server and cloud server).
- All the participants have some public and also some private information and only the appropriate participants can have access to the private information. To maintain the level of abstraction smart contract is used.
- Proposed system is tested in a real world platform as a proof-of-concept and discussed storage, communication and computational costs.

In Sect. 2, ambulance management related systems and the effect of edge computing in healthcare sector are discussed with the motivation of the proposed work. The proposed system structure is discussed with all its components and transactions in Sect. 3. Implementation details are provided in Sect. 4 and performance analysis of the system is presented in Sect. 5. At the end, conclusion of the paper and the future research opportunities are briefly discussed in Sect. 6.

2 Related Works

The related work section is divided into three parts. In the first part, contributions of edge computing in the area of healthcare will be presented. The utilization of blockchain in healthcare as well as in the field of ambulance management will be discussed in the second part and finally motivation of this work will be explained at the end.

2.1 Edge Computing in Healthcare

Although the idea of edge computing is not new, usages of increased a lot after increasing the popularity of light-weight IoT devices. As healthcare sectors are taking advantages from different types of IoT devices, edge computing is there to provide computation support. For example, as it is distinct from the already existent offloading frameworks, health monitoring of patients is exceedingly delay-sensitive. Thus, with the underlying delay constraint of remote cloud servers, satisfactory services are not always received. To reduce the transmission latency, edge computing can be utilized as a solution. Pace et al. in [5] propose utilizing hybrid cloud computing to preserve privacy, as there is privacy issue in edge computing based health monitoring systems. The authors in [6] describe a cost efficient healthcare system with integrated edge computing and health monitoring. The proposed framework depicts the association of servers, distributed medical analysis tasks and with virtual machines deployed. For the solution of the optimization problem, linear programming based, heuristic method is developed [2] proposes the window-based Rate Control Algorithm (w-RCA) for the optimization of the medical quality of service (m-QoS) in the mobile edge computing enabled healthcare by considering the network parameters peak-to-mean ratio (PMR), standard deviation (Std.dev), delay and jitter during 8 min medical video stream transmission over 5G networks. These papers have shown the usage of mobile edge computing in different areas of healthcare with distinct integrated frameworks. From our research we are yet to find a smart ambulance service with blockchain integrated edge computing which will preserve privacy as well as solve the delay-sensitive issue for monitoring health.

As mentioned above some applications of edge computing has been found in the healthcare sector but unfortunately there is no work related to ambulance management.

2.2 Blockchain in Healthcare

Utilization of blockchain in healthcare is not new. For example, Dubovitskaya et al. [7] presented a framework for remote patients from different locations and medical departments. A blockchain is utilized to store and manage online medical records that are shared between patients and healthcare organizations to prescribe medicines. The frameware also provides access to doctors and specialized experts upon request. Similarly, in [8, 9] blockchain integrated frameworks were presented where cloud is used to store and monitor patients' healthcare records to the authorized entities where Hossein et al. in [10] and Grishin et al. in [11] proposed similar structure by using blockchain with additional privacy preserving facilities. In [12], a blockchain integrated medical service providing approach is demonstrated where cloud server is used to store and share treatment histories and diagnosis reports. Luo et al. [13] has come up with a blockchain framework with an additional levels of privacy protection facility by imposing two level authentication that are general and role-based authentication. Ferrag et al. used blockchain to mange privacy preserving data transmission between IoT devices used for healthcare in [14].

On the other hand, generally ambulance management is performed by phone call, but now it is required to think about online ambulance management system specially because of the current pandemic situation where providing instant ambulance service is very hard to achieve. Utilization of blockchains in the area of ambulance management is not new though. For example, Nkenyereye et al. proposed an emergency driven message (EDM) protocol based on blockchain which record EDMs in a distributed system for 5G-enabled vehicular edge computing in a private blockchain along with edge nodes. For fast and efficient movement of ambulances to the place of the accident, EDMs are sent [15]. Akhter et al. [16, 17] proposed an authentication protocol for VANET system with blockchain where emergency vehicles like ambulances are given priority based services to reach to the accident place or hospital faster. Several projects for smart ambulances have been proposed with additional features and facilities for healthcare. For example, a smart ambulance integrated with all necessary kinds of sensors to monitor vital signs and other necessary datas, was proposed by Udwant et al. These data will be transmitted to the hospital's database and via GPRS message traffic signals will be operated but here the important patient data don't have much security as it is saved in the cloud. By using IoT scenario, the traffic can be cleared by sending messages to the signal board to make the ambulances reach the destination faster by making use of Embedded and IoT, a model is introduced [18] An IoT based live system is introduced to monitor patients at risk. This sends hospital alarming notification if the patients get into critical condition. Furthermore a hardware integrated ambulance with live trafficking system is introduced for quicker and safer transit respectively [19].

2.3 Motivations

Although ambulance management is a very important part of healthcare system, often this part is ignored because of its simplicity. But in the current pandemic situation to handle huge pressure of patients special features are required to be added to increase the efficiency of ambulance management. Thus, in this papers an ambulance management system is presented by combining the latest technologies together.

All the above mentioned papers have shown ambulance related services, blockchain based healthcare systems or applications of edge computing. Our research has yet to lead use to a paper with blockchain integrated ambulance service from hospitals neither could we find edge computing servers used to make computational work efficient. Proposed ambulance management system can increase efficiency specially for the patient from rural areas. With video conversation with doctors and primary diagnosis facilities ambulances will act as a small medical center which is live saving. With the help of blockchain, patient sensitive data management become easy and more efficient in terms of security and other facilities.

To manage huge data load and computational support for the proposed ambulance management systems edge computing and cloud is used. This will remove the infrastructural and maintenance cost and it is possible to get world class security services for the health data. Patients are required to give their sensitive medical history or tests data to medical databases to keep their own health records. These need to be secured, decentralized, robust, flexible, immutable with privacy preserving support and can only be delivered to the appropriate service providers i.e., medical centers, doctors, etc. which can be managed by blockchain efficiently [17]. For big amount of data transmission and video streaming it requires high speed internet and currently available 5G internet will be enough to provide a smooth communication between the participants of the system. All the information of the participants (patients, ambulances, doctors and medical centers) will be stored in the blockchain with their activities as transactions. Participants public information will be available to all where private information will be available to only to the applicable participants. The level of abstraction is created by using smart contract. Blockchain transactions required high computational power, thus in the proposed system edge server will be used by the participants to handle all the computational tasks. Similarly, huge amount of data may required to be stored and thus cloud service is used.

3   System Structure

A smart ambulance management system for hospitals is proposed here where ambulances are equipped with primary investigation and video transmission facilities. A blockchain is used where patients, ambulances of medical centers are the member nodes. Members are connected through a high-speed internet connection and all their activities will be considered as blockchain transactions to ensure authenticity, non-repudiation, integrity, confidentiality, security, attack prevention facilities. To ensure the availability of the services, edge computing is used to process the transactions and all other complex computational tasks required for the system so that a device with minimum computation power can participate. Figure 2 illustrates the components of the system and details will be discussed in this section.

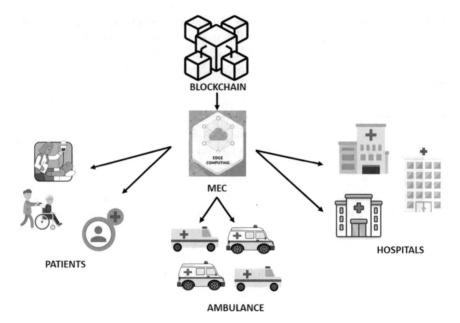

Fig. 2 System structure

3.1 Participants of the System

There are four main participants of the system. Those are: patients, ambulances, doctors and medical centers. Among the participants ambulances, doctors and medical centers will be called as service providers while patients will be considered as customers. All the participants have to register to the system with their basic information and the service providers have to provide a list of additional features and services. All the information will be stored in the blockchain and the participants are considered as entities of the blockchain. Among the components of the system some information will be considered as public which will be available for all the participants. On the other hand private information like patients' details, diagnosis information, account credentials, communication information will be considered as private. The level of abstraction is ensured by using smart contract.

3.2 Transactions

Activities of the components will be considered as blockchain transactions. For the proposed system, list of transactions are the followings:

1. Advertisement of features by service providers after registration
2. Request for service by patients (ambulance, doctor, medical center)
3. Response from service provider
4. Response by patient to a specific service provider
5. Acceptance of a service
6. Exchange of credentials to access the cloud (to store diagnosis, results, etc.)
7. Exchange of credentials for video conferencing
8. Completion of a service
9. Reputation scores submission by customers.

3.3   Working Procedure

A patient can register to the system and ask for a service like ambulance, medical center or doctors' appointment. Edge server receives the request and verify authentication information of the sender. A block is created with that transaction and broadcast to the nearby service providers. The area is decided according the service providers range of service which was previously given by them. Upon receiving a service request interested service providers send a response through the system which includes the list of available time slots, locations, features, service charge, insurance information, etc. Patient can check the list of available services and have the freedom to select any of the services. After receiving the response from the patient, edge server creates another block in the blockchain and allocate a space for this agreement and share necessary information which includes credentials to access the cloud and video conferencing, diagnosis report form, etc. Ambulance or medical centers may have remote doctors' support facility. In that case, if patients want that service the condition of the patient can be monitored full time during the transit period and can provide necessary instructions which can be life saving. To minimize the storage consumption, blockchain is not involved in the video conferencing and diagnosis information storage part, rather it just share the credential information between the participants. The system ensures the privacy of the patients' personal and diagnosis information by make them available only to the related parties. Moreover, inside the system all the components are known by their public keys rather than real identity to preserve their privacy. To ensure the quality of service, after completion of an agreement patients can provide their feedback about the service provides which will be available to all the customers.

A flow chart is presented in Fig. 3, to demonstrate the complete working procedure of the system.

3.3.1   Data Transmission Details

Whenever a patient agreed to get service from a particular service provider, the edge sever will store that as a transaction. After that, the system will provide some spaces

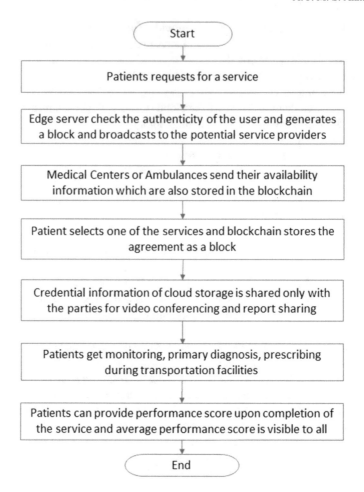

Fig. 3 Flow chart to demonstrate the working procedure of the proposed system

in the cloud so that the service providers can upload necessary data into that place which includes diagnosis information, video conferencing related information, etc. Typically, smart ambulances have the capability to measure temperature, blood pressure, pulse rate, oxygen saturation level, respiration rate, ECG, EEG, ultrasound video, etc. It requires less than 10 kbps of speed to transmit most of the information while for electrocardiography (ECG) and electroencephalography (EEG) 90+ kbps transmission speed is expected [20]. After measuring those, attendant in the ambulance can upload those into the cloud space provided by the system. That space is only available to the patient, ambulance, medical center and the doctor who can provide live support to ensure the privacy of the patients' information. First of all, COVID-19 related tests can be performed and if the results come positive the patient can ask for specialized hospitals for COVID patients in the system. Primary diagnosis report also helps the medical centers' authority to allocate specialized doctor required for a particular patient.

3.3.2 Video Calling Service

To perform video conferencing, both the ambulance and medical centers need high speed internet connections (5G or above) and HD video capturing capability. Doctors from medical centers or even from home can provide emergency supports during the transportation period and responsible attendant in the ambulance can take necessary steps which can be proved life saving specially for the patient coming from remote areas. Moreover, instant diagnosis services available in the ambulance will help the doctor to provide more effective supports. Credentials from video conferencing will be provided by the edge server which will be exclusive to that particular transportation only to ensure the privacy and security of these information.

4 Implementation

The proposed system is implemented in three phases. Firstly, the front end of the system is developed by using html and css. Secondly, a virtual machine is prepared as edge server where a blockchain simulator is hosted. Finally, four more virtual machines are prepared to represent patient, ambulance, medical center and remote doctor. To manage the transactions, block generation processes are written in solidity programming language and deployed as smart contract. Privacy of the information is also managed by the smart contract.

Oracle VM VirtualBox 6.1 is used to simulate edge server and the participants. Configuration of the virtual machines are presented in Table 1. Truffle framework [21] is used to implement the blockchain as it provides necessary tools to compile and deploy smart contracts, built in blockchain emulator named ganache [22] with customizing, logging and debugging capability. Node packet manager (NPM) [23] is used to run JavaScrips and lightweight node server [24] is used to enable web services. Participants of the system use metamask [25] wallets to connect with the system by using web browsers.

The list of functions used to run the system are presented in Table 2. To test the system, after deploying the blockchain, all the participants established their connections with the server VM and perform registration processes by providing necessary information. Then, the Patient-VM sends a request for ambulance which will be processed by request_for_ambulance function. A response from the ambulance-VM is generated which will be functioned by response_from_service_providers function. The patient accept request of the responded ambulance and the system handle it by using accept_from_responses function. Similarly, same procedure is repeated for doctor and medical centers.

Table 1 Implementation parameters for blockchain based authentication

Machine	CPU	Memory (GB)	Storage (GB)	OS
SERVER-VM	6	8	60	Ubuntu-18.04
Patient-VM	1	2	20	Ubuntu-18.04
Ambulance-VM	1	2	20	Windows 7
Doctor-VM	1	2	20	Ubuntu-18.04
Medical-VM	1	2	20	Windows 7

Table 2 Functions to handle blockchain transactions (Patients, ambulance, medical centers and doctors are represented by P, A, D and MC respectively)

Function	Performed by			
Name	P	A	D	MC
registration()	✓	✓	✓	✓
request_for_ambulance()	✓	–	–	–
request_for_medical()	✓	–	–	–
request_for_doctor()	✓	–	–	–
response_from_service_providers()	✓	✓	–	✓
accept_from_responses()	✓	–	–	–
credentials_for_storage()	✓	–	–	–
credentials_for_video_monitoring()	–	–	–	–
completion()	✓	✓	✓	✓
termination()	✓	✓	✓	✓
rating()	✓	–	–	–

After the agreement, the edge server allocate some spaces for the service providers to drop necessary information like diagnosis info, video monitoring data, etc. and sends it by using credentials_for_video_monitoring function. Termination, completion and service rating functions are also tested during the experiments.

5 Performance Analysis

In this section, performance of the proposed method is explained in two parts. Firstly, storage and computational time requirement of the implemented blockchain based system is presented and then data rate, latency and packet loss ratio during information transmission is explained.

5.1   Performance of Blockchain Based System

5.1.1   Storage Requirement

In ethereum blockchain block headers consume 508 bytes and because of encryption, it acquires more than 1024 bits (128 bytes) which make a total of 636 bytes per transaction [26]. Thus even if ten thousand blocks are generated every day it acquires only 6 MB of data that means 2.1 GB per year which is a minimum requirement to ensure security, privacy, confidentiality, integrity, etc. of the data.

5.1.2   Execution Time

Block generation is a time-consuming job, thus none of the components will perform block generation rather they just send the transaction to the blockchain server. The server will be responsible to generate blocks from authorized entities and broadcast them to the nearby service providers. The server performs instant mining and requires ignorable time.

The components use a lightweight digital signature algorithm called RSA-1024. It provides a security strength of 80-bit which means it required at least 2^{80} operations to break the private key and is considered as pretty good security for devices with low computational power. By using this, it requires 1.48 ms to sign a message and 0.07 ms to verify that, thus only 1.55 ms is required to complete one transaction for a computer with only 1.5 GHz processing speed [27]. Other blockchain-based systems takes 10–20.1 ms time ([28–33]) who use different method required more time than the proposed method [16].

5.2   Performance of Data Transmission

By using the above mentioned experimental setup, diagnosis data and video calling related information are uploaded by the ambulance in a cloud storage and later accessed by the medical center and patient. The transmission performance will be described in this section in terms of latency, data rate and packet loss ratio. The experimental data includes tele-medicine and tele-surgery are collected from [34].

5.2.1   Transmission Delay

Figure 4 depicts the different requirements for our data transmission. In a two way communication system, the maximum allowed latency for a two-way communication for audio-visual data is the same which is 150 ms. Whereas, for vital signs it is slightly higher (250 ms). For the 5G enabled connected ambulance, having eMBB and URLLC these requirements are met.

Fig. 4 Data transmission
delay

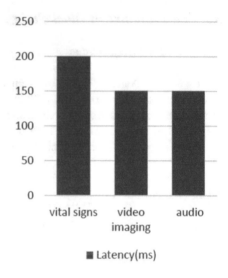

■ Latency(ms)

5.2.2 Data Rate

Various data types require different data rates for the transmission and reception to
be seamless. Observed results in Fig. 5 show that the highest data rate requirement
is for two-way visual multimedia streaming with 10 Mbps, followed by the audio
multimedia stream with a requirement of 200 Kbps. Depending on the required
quality and bandwidth constraints the data rate requirements for audio data range
between 22 and 200 Kbps. Different types of vital signs require different data rates
that are outlined in Table 3, with EEG requiring up to 86.4 Kbps.

Fig. 5 Data rate

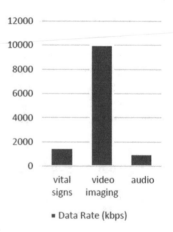

■ Data Rate (kbps)

Table 3 Required data rate for diagnosis data

Data type	Data rate (kbps)
Blood pressure	< 10
Pulse rate	<10
Temperature	<10
Respiration rate	<10
Oxygen saturation	<10
ECG	72
EEG	86.4

5.2.3 Packet Loss Ratio

Minimum data loss is required for ultra-reliability. Estimating the channel quality and consequently choosing the best modulation and coding scheme (MCS). In a 4G network, channel quality indicator (CQI), ranging between 1 and 15, is used to assign the best MCS. The same method is expected to be used for 5G networks with polar coding and LDPC coding joining or replacing Turbo and convolutional coding. Higher order modulation schemes are also expected to join the CQI table along with new and robust channel coding schemes. Observing the results in Fig. 5c, depicts requirements for PLR are slightly lower two way audio flow but same for visual multimedia streaming and vital signs (Fig. 6).

Fig. 6 Packet loss ratio

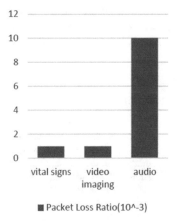

■ Packet Loss Ratio(10^-3)

5.3 Security Services

By using blockchain with RSA-1024 digital signature algorithm the proposed method ensures the security of 80-bit strength, integrity and confidentiality of the transmitted data.

Because of the signing method only authenticated members are allowed to perform a transaction and it also ensures non-repudiation of the members.

By using public keys to communicate with each other the system preserves the privacy of the patients' information. Additionally, a level of abstraction is created by using smart contracts so that only the corresponding members get access to the patients' diagnosis information.

Blockchain is utilized to provide a decentralized and distributed environment to establish a communication platform between patients with ambulances and medical centers. Blockchain also provides additional services like temper-resistance, immutability, transparency, fairness and robustness of the data.

6 Conclusion

From the best of our knowledge, a ambulance management system managed by blockchain and supported by edge computing and cloud is not available. All the previous works either use blockchain to relay messages or emergency-driven messages to decrease transit time by relaying messages to control traffic or they are traditional in ambulance treatment related papers. The use of blockchain has been seen in healthcare related papers to secure patient sensitive data. This work uses blockchain to secure the participants sensitive information by using smart contract, managed by blockchain and supported by edge computing and cloud.

References

1. Longo DL, Drazen JM (2016) Data sharing
2. Sodhro AH, Luo Z, Sangaiah AK, Baik SW (2019) Mobile edge computing based qos optimization in medical healthcare applications. Int J Inf Manage 45:308–318
3. Akhter A, Ahmed M, Shah A, Anwar A, Kayes A, Zengin A (2021) A blockchain-based authentication protocol for cooperative vehicular ad hoc network. Sensors 21(4):1273
4. Akhter A, Shah A, Ahmed M, Moustafa N, Cavusoglu U, Zengin A (2021) A secured message transmission protocol for vehicular ad hoc networks. CMC-Comput Mater Contin 68(1):229–246
5. Pace P, Aloi G, Gravina R, Caliciuri G, Fortino G, Liotta A (2018) An edge-based architecture to support efficient applications for healthcare industry 4.0. IEEE Trans Ind Inf 15(1):481–489
6. Gu L, Zeng D, Guo S, Barnawi A, Xiang Y (2015) Cost efficient resource management in fog computing supported medical cyber-physical system. IEEE Trans Emerg Top Comput 5(1):108–119

7. Dubovitskaya A, Xu Z, Ryu S, Schumacher M, Wang F (2017) Secure and trustable electronic medical records sharing using blockchain. In: AMIA annual symposium proceedings, American Medical Informatics Association, vol 2017, p 650
8. Kaur H, Alam MA, Jameel R, Mourya AK, Chang V (2018) A proposed solution and future direction for blockchain-based heterogeneous medicare data in cloud environment. J Med Syst 42(8):1–11
9. Khezr S, Moniruzzaman M, Yassine A, Benlamri R (2019) Blockchain technology in healthcare: A comprehensive review and directions for future research. Appl Sci 9(9):1736
10. Mohammad Hossein K, Esmaeili M, Dargahi T, Khonsari A, et al (2019) Blockchain-based privacy-preserving healthcare architecture. In: 2019 IEEE Canadian conference of electrical and computer engineering (CCECE), IEEE
11. Grishin D, Obbad K, Estep P, Quinn K, Zaranek SW, Zaranek AW, Vandewege W, Clegg T, César N, Cifric M et al (2018) Accelerating genomic data generation and facilitating genomic data access using decentralization, privacy-preserving technologies and equitable compensation. Blockchain in Healthcare Today 1:1–23
12. Xia Q, Sifah EB, Asamoah KO, Gao J, Du X, Guizani M (2017) Medshare: trust-less medical data sharing among cloud service providers via blockchain. IEEE Access 5:14757–14767
13. Luo Y, Jin H, Li P (2019) A blockchain future for secure clinical data sharing: a position paper. In: Proceedings of the ACM international workshop on security in software defined networks & network function virtualization, pp 23–27
14. Ferrag MA, Derdour M, Mukherjee M, Derhab A, Maglaras L, Janicke H (2018) Blockchain technologies for the internet of things: research issues and challenges. IEEE Internet Things J 6(2):2188–2204
15. Nkenyereye L, Adhi Tama B, Shahzad MK, Choi YH (2020) Secure and blockchain-based emergency driven message protocol for 5g enabled vehicular edge computing. Sensors 20(1):154
16. Akhter A, Ahmed M, Shah A, Anwar A, Kayes A, Zengin A (2021) A blockchain-based authentication protocol for cooperative vehicular ad hoc network. Sensors 21(4):1273
17. Akhter A, Ahmed M, Shah A, Anwar A, Zengin A (2021) A secured privacy-preserving multilevel blockchain framework for cluster based vanet. Sustainability 13(1):400
18. Hölbl M, Kompara M, Kamišalić A, Nemec Zlatolas L (2018) A systematic review of the use of blockchain in healthcare. Symmetry 10(10):470
19. Saha HN, Raun NF, Saha M (2017) Monitoring patient's health with smart ambulance system using internet of things (iots). In: 2017 8th annual industrial automation and electromechanical engineering conference (IEMECON). IEEE, pp 91–95
20. Usman MA, Philip NY, Politis C (2019) 5g enabled mobile healthcare for ambulances. In: 2019 IEEE Globecom workshops (GC Wkshps). IEEE, pp 1–6
21. Truffle suite. https://www.trufflesuite.com/. Accessed from 2020-04-08
22. Ganache. https://www.trufflesuite.com/ganache
23. Npm (software). https://en.wikipedia.org/wiki/Npm_software. Accessed from 2020-04-08
24. Github lightweight node server. https://github.com/johnpapa/lite-servers. Accessed from 2020-04-08
25. Metamask. https://metamask.io/. Accessed from 2020-04-08
26. Akhter A, Ahmed M, Shah A, Anwar A, Zengin A (2021) A secured privacy-preserving multilevel blockchain framework for cluster based vanet. sustainability 2021, 13, 400
27. Nirala RK, Ansari MD (2018) Performance evaluation of loss packet percentage for asymmetric key cryptography in vanet. In: 2018 Fifth International Conference on Parallel. Distributed and Grid Computing (PDGC). IEEE, pp 70–74
28. Azees M, Vijayakumar P, Deboarh LJ (2017) Eaap: efficient anonymous authentication with conditional privacy-preserving scheme for vehicular ad hoc networks. IEEE Trans Intell Transp Syst 18(9):2467–2476
29. Rongxing L, Xiaodong L, Xuemin S (2010) Spring: A social-based privacy-preserving packet forwarding protocol for vehicular delay tolerant networks. In: Proceedings of IEEE INFOCOM, pp 1–9

30. Shao J, Lin X, Lu R, Zuo C (2015) A threshold anonymous authentication protocol for vanets. IEEE Trans Veh Technol 65(3):1711–1720
31. Wang C, Shen J, Lai JF, Liu J (2020) B-tsca: Blockchain assisted trustworthiness scalable computation for v2i authentication in vanets. IEEE Trans. Emerg. Top. Comput
32. Zhang C, Lu R, Lin X, Ho PH, Shen X (2008) An efficient identity-based batch verification scheme for vehicular sensor networks. In: IEEE INFOCOM 2008-the 27th conference on computer communications. IEEE, pp 246–250
33. Zhang X, Chen X (2019) Data security sharing and storage based on a consortium blockchain in a vehicular ad-hoc network. IEEE Access 7:58241–58254
34. Zhang Q, Liu J, Zhao G (2018) Towards 5g enabled tactile robotic telesurgery. arXiv:1803.03586

Cryptocurrency: Is it the Future of Payments?

Zachary Mineau, Dylan Hoffman, Jonathan Lor, and Nazim Choudhury

Abstract Cryptocurrency, which is built using blockchain technology, is one of the newest methods of doing secure transactions without a central authority. Due to the unregulated means in which cryptocurrency is transferred, it becomes incredibly hard to ensure the fairness and legitimacy of the transactions. This is to further explore the history, process, problems, competition, advantages, and the current future of cryptocurrency. Finally, the team built its cryptocurrency and blockchain similar in functionality to Bitcoin, which is currently the most popular cryptocurrency by sheer market capitalization of over \$898 Billion. This allows people to track the flow of cryptocurrency in all its flaws, and its potential succession of other currencies. This chapter explore different aspects of the cryptocurrency which is a vital aspect of futuristic cyber smart cities. This chapter will serve as a reference for cyber smart city developers and other relevant stakeholders in designing futuristic cities.

1 Introduction

Currency is the fundamental essence of trade in the modern world. It is the legal tender for which people, businesses, and governments recognize and accept payment for various goods and services. With the advancement of fiat currencies, the value of the currency is not in the physical item, but rather the fundamental idea that the item is worth something more. This is the basis for all cryptocurrencies. It stands as a self-regulated virtual fiat currency that utilizes internet access. Therefore, it is important to understand the history, problems, competition, advantages, the future, and the process of how cryptocurrencies work.

The original cryptocurrency that struck out big was Bitcoin. Bitcoin was launched in 2009 by a user unknown. The only identification is a pseudonym, Satoshi Nakamoto. The promotional material for the new currency was that cryptocurrencies did not rely on trust of an organized banking system. Instead, it utilized "Crypto-

Z. Mineau · D. Hoffman · J. Lor · N. Choudhury (✉)
Department of Computer Science, University of Wisconsin Green Bay, Green Bay, USA
e-mail: choudhun@uwgb.edu

© The Author(s), under exclusive license to Springer Nature Switzerland AG 2023
M. Ahmed and P. Haskell-Dowland (eds.), *Cybersecurity for Smart Cities*,
Advanced Sciences and Technologies for Security Applications,
https://doi.org/10.1007/978-3-031-24946-4_12

graphic proof rather than trust" [1]. This prompted one of the greatest forms of tracking for online currency. Although Bitcoin was created and nuanced in 2008, it did not get worldwide attention until 2011.

Since Bitcoin utilizes proof of independent users, it led to many problems. One of these problems was that Bitcoin had no central authority. This meant that the ownership of Bitcoin had to be continuously monitored. A study in 2017 that showed the total mining revenue per year. The revenue gained by bitcoin mining is changing year to year. This is one of many reasons that people are hesitant to back bitcoin [2]. The revenue gained from mining is inconsistent at best. There is also a lack of immediate data from miners.

Another major problem is that cryptocurrencies, especially Bitcoin, are affected by the perception of the value that they hold. Data was collected on several cryptocurrencies to determine how each currency value rose or fell to outside effects and inside effects with each other. After several months it was shown that the overall pricing of the currencies was affected by the belief of the currency. As the media talked about a currency, the price of that currency would go up. When a negative story was published or no activity was published, the price would go down. It is suggested that it would be unlikely to sustain prices of cryptocurrencies as a financial asset if no one believed in its potential [3]. This poses a problem with any fiat currency.

Still, not all fiat currencies share all weaknesses. Cryptocurrencies are liable to DDos attacks. Some miners have shown that a DDoS attack can cause a system malfunction [4]. With every successful attack, the developers continue to update and perform bug fixes. However, there have been cases where a bug fix was not successfully updated at an exchange firm that led to an attack. The Magic the Gathering Online Exchange firm was hit when a hacker was able to rob them of approximately 460 million USD. It was found out that the CEO of the firm was not updating the systems regularly. Since the system was not updated, it ended up costing hundreds of millions of dollars [5]. This is a major concern to any company or government.

1.1 Chapter Roadmap

Rest of the chapter is organized as follows. Section 2 discusses the government recognition for cryptocurrency. Section 3 showcases the key concerns, Sect. 4 discusses the competition for the cryptocurrency, Sect. 5 highlights the advantages, Sect. 6 showcases the cryptocurrency processing and finally the chapter is concluded in Sect. 7.

2 Government Recognition

When a country does not recognize a currency, it causes numerous problems. The main restriction to adopting governmental backing is that the government must impose regulations upon the currency. With regulations, the currencies that are struggling to obtain acceptance will gain greater legitimacy in the larger audience [6]. Inside Bitcoin, however, due to virtual currency being in the infancy stage, it is likely that many changes will occur before a nation backs any cryptocurrency.

If companies and countries begin to accept cryptocurrency as a legit currency, it will only ensure the reliability and sustainability of bitcoin. With the acceptance of the currency, many opportunities become available. It has been noted that bitcoin is great for international transactions. Although money can be wired to another country already, there are hurdles. Different countries have different values which means the money must be converted. Some transactions can elicit fees and even denial of service. If money needs to be transferred quickly to another country, having a unanimous currency that is unrestricted by governance is the way to go.

Some countries, like Canada, have already begun adopting virtual currency [7]. The system seeks to minimize the risks associated with cryptocurrencies by imposing the before mentioned regulations. This further benefits cryptocurrency as the Bank of Canada has acknowledged the developing virtual currency market. However, some countries like Russia are concerned with the emergence of virtual currency. The Bank of Russia has stated that it violates federal law stating that all currency must go through a central bank. Another concern from Russia is that the value of the coin fluctuates every day. It is nearly impossible to set a standard for the coin as it changes without proper regulation.

3 Major Concerns

This section discusses the major issues in adopting cryptocurrency in the futuristic context, such as smart city perspective.

3.1 Frauds and Scams

Even if all the bugs in the systems get patched and every country recognizes cryptocurrencies, there are always going to be problems. The ledger and block chain are public which allows for semi-anonymity [8]. A problem that can never be truly solved is scams. With the evolution of technology protection, the ways around that security also evolve. People are using scams, ransomware, Ponzi schemes, and most of these schemes can be traced to links associated with social media.

Scams have always been a part of society. If there has been coin and product to swindle, a swindler will appear. It is no different with cryptocurrencies. These scams tend to fall under two categories. The first is to convince the victim to transfer cryptocurrency to a blockchain address with a promise of returning more cryptocurrency. The old spend money to make money schemes. The second category revolves around the victim providing existing credentials that are required to access their private cryptocurrency account with the promise of additional funds. Thankfully, as these scams get noticed, people warn others that it exists, and this means that the scammers resort to other tactics.

Another one of the tactics is ransomware. This is where a hacker gains access to a victim's computer system or other personal information. The hacker then threatens the victim with sharing, using, or even deleting all personal information accessible. A price is then set for an amount of a cryptocurrency to be sent to the hacker [9]. Cryptocurrency is great to have as a means of criminality as it is hard to trace where the account is and who truly owns that account. This is a problem because there is no central bank that runs all the accounts with a registered person behind each account. However, these are still able to be monitored by the blockchain.

The next tactic is simple Ponzi and pyramid schemes. These scams claim to be verified and trustworthy organizations that promote cryptocurrency. Some creators and influencers create their own version of a cryptocurrency. They hype it up to their audience and have them buy into the crypto. Once these initial people buy up the coins, the original creator leaves and the victims then must rely on selling the coin to someone else to regain what they lost [10]. Since the coin does not function as legal tender, and newcomers are slowly dwindling, those who bought the coin are without any reimbursement.

All these tactics are different, but they do have something in common. Most of the various versions of these tactics utilize social media to target and link each account to each other. It is common to see famous people like Elon Musk offering free cryptocurrency. However, these famous people are typically run by imposters [11]. This occurs when a user creates a fake Twitter or other social media account and pretends to be a well-known individual. Since people like Elon Musk are known for their technology savvy ways, people believe that the free cryptocurrency is real. The only real comfort comes in the way the scams are usually shut down quick. In fact, most of them that are recorded are no longer functioning [12]. When they click on the link they can easily fall into a scam.

3.2 Scalability Concerns

One problem of proof-of-work consensus is its scalability. Miners must verify each and every transaction on the network. As cryptocurrencies continue to gain popularity and more people join the network, the number of transactions increases as well. However, because every transaction must be validated sequentially, more steps are required for the transaction to be verified. It currently takes anywhere from 10 min to one hour to verify a Bitcoin transaction. Bitcoin verifies transactions at a max of seven transactions a second, but realistically verifies less than that. For reference, PayPal can verify around 193 transactions per second while Visa is able to verify around 1700 transactions a second. Ethereum also shares this scalability issue with Bitcoin. Ethereum on average processes around 20–25 transactions a second, way down from its theoretical 1000 transactions per second [13]. This is due to Ethereum's transaction fees, also known as "gas", a fee determined by the "amount of computational effort required to execute specific operations on the Ethereum network". If Bitcoin or Ethereum are to become staples in the global economy, the relatively slow transaction verification becomes a big problem.

3.3 Energy Concerns

On March 24, 2021, Elon Musk, Tesla CEO tweeted, "You can now buy a Tesla with Bitcoin". Just 49 days later, on May 12, 2021, Elon Musk tweeted "Tesla & Bitcoin", alongside a statement that Tesla would be suspending vehicle purchases with Bitcoin [14]. Musk cited fossil fuel and energy consumption concerns regarding Bitcoin mining and transactions as the cause of the suspension. The statement also included praise for cryptocurrency in general and believes it has a promising future but was skeptical about the energy consumption of cryptocurrency. In a New York Times interactive article about Bitcoin, it's estimated that just Bitcoin itself consumes about 91 terawatt-hours of electricity annually, more than the country of Finland. Bitcoin's electricity usage has increased tenfold in about five years [15].

Proof-of-work is the most popular consensus protocol currently used in the cryptocurrency environment and a vast majority of cryptocurrencies use proof-of-work. Bitcoin and Ethereum, the two most popular cryptocurrencies both currently use proof-of-work consensus. Proof-of-work is a consensus mechanism wherein users or "miners" compete to solve an arbitrary math problem to become the first to validate the next block. The miner that solves the math problem first and validates the next block is awarded many coins. Miners, whether individuals or companies, compete to validate transactions to get coin rewards. The more computer power you have, the more likely you are to solve this arbitrary math problem and be rewarded with coins. As the price of cryptocurrencies rises, more and more miners are incentivized to join. As more miners participate in mining, the math problem becomes more difficult, requiring more and more energy to solve.

A study compared the US energy consumption to the trading volume of cryptocurrency, from 2014 to the end of 2017 [16]. The study found a positive correlation between the trading volume of cryptocurrency and energy consumption. As the trading volume of all cryptocurrencies went up, energy consumption in the US also went up. In regard to Bitcoin, the study concluded that the "trading of bitcoin appears to have a long-run positive influence on the production of energy" and that this energy consumption growth is a limitation of Bitcoin in terms of sustainability. As Bitcoin is the poster child of cryptocurrencies, Bitcoin's price and sentiment has a significant impact on the price and sentiment of other cryptocurrencies, which in turn incentivizes more miners to partake, thereby increasing energy consumption.

The rise of cryptocurrency prices over the decade has incentivized crypto miners to build data centers for the sole purpose of mining cryptocurrency. These data centers have a high demand for power, so miners are placing data centers in location with relatively cheap energy costs. The Greenburg and Bugden study [17] studied Chelan

County, Washington, where its abundance of cheap power attracted an influx of crypto mining to the community beginning in the early 2010s. This influx created an "energy consumption boomtown", and how the county is dealing with the increase in energy consumption over the years as Bitcoin and other cryptocurrencies grew.

In 2014, the county enacted a moratorium after receiving 34 power inquiries for the use of 220 MW of electricity, double the energy use for the entire county. In 2017, after another explosive Bitcoin price growth, Chelan County saw energy consumption considerably increase and enacted another moratorium on miners. In 2018, Wenatchee, the largest city in the county, banned crypto mining in residential and mixed-use areas for a year due to unauthorized miners overtaxing the power system. As Bitcoin and cryptocurrency continue to grow, increase in price, and gain popularity, more energy consumption boomtowns are likely to show up around the world.

4 Cryptocurrency Competition

After Bitcoin took off, it was not long before other people latched onto the cryptocurrency idea and changed it to how they would prefer it. This sparks competition between the cryptocurrencies. These competitors are called Altcoins [18]. All Altcoins take portions of Bitcoin and mimic it by creating slight differences in the structure. Since all cryptocurrencies utilize the blockchain model from Bitcoin, each cryptocurrency takes that section of code. Blockchain is the common statistic between them all. How that blockchain is used and monitored is where the differences come into play. As new currencies appear, new methods of reading the blockchain and monitoring transactions are created as well. This is the power of Altcoins.

4.1 Altcoins

Most Altcoins are like Bitcoin and build upon preexisting code. The best example of this is that most Altcoins use blockchain methods. To show this, an example of an Altcoin is Litecoin. A major promotion of Litecoin is that it can generate four times as many coins as well as add the transactions to the blockchain four times faster [19]. Another example of an altcoin making significant changes is Peercoin. Like how bitcoin utilizes blockchain and proof of work, Peercoin takes these efforts and improves upon them. Peercoin utilizes proof of stake alongside proof of work. With this, Peercoin can mitigate the need for powerful, and expensive computers for mining [20]. This makes it more acceptable to a common household family as a viable currency.

Bitcoin is the leading form of cryptocurrency. In 2017 the revenue of bitcoin was drastically higher than its competitors. Bitcoin made over $25 billion while other cryptocurrencies made just over $5 billion [21]. Litecoin and Peercoin are the next leading options for cryptocurrency. Litecoin in 2011 and Peercoin in 2012. The purpose of the cryptocurrencies is not solely in competition, but to solve some of the inherent problems Bitcoin caused. As interest in cryptocurrency grows, so does the interest in creating a different currency. This is where the root of the competition begins.

5   Advantages of Cryptocurrency

The issues with the main system for doing payments of the internet i.e., giving our credit card information to a seller, is that people are exposing our identity as well as sensitive credit card information to the seller, let's say Amazon, and then that information is passed to another financial system, let's say MasterCard. Not only does this expose our financial information, but it also provides no anonymity when doing transitions online. One potential solution to this problem would be an intermediary between buyer and seller, such as PayPal. Here, both buyer and seller have an account with PayPal.

When a buyer wants to do a transaction with a seller, all they need to do is tell PayPal to charge the buyer for the transaction [22]. Since PayPal has both the seller's payment information and buyer's, they can just directly charge the buyer and they credit the seller without giving any of the payment details to the seller directly. While this is an improvement to previously giving our payment information directly to a seller, there is still the problem of anonymity that is not solved. A PayPal account is tied directly to a person that can be easily traced. Another problem with this system is there is a central authority, that is PayPal that is governing over the transactions, much like the banks do today.

Another failed system that predates Bitcoin was DigiCash. DigiCash, which was patented by David Chaum sought to seek out the challenges of creating a digital coin that deals with the classic double spend problem. Chaum's implementation uses what is called a "blind signature" where a central authority issues a coin, but you get to pick the serial number of it, then the central authority signs it, this way the coin is verifiable as being spent or not. DigiCash also relied on merchants to support it. With DigiCash, buyers are anonymous, but the merchants are not. When doing a transaction on a merchant's website, the DigiCash system would open a secure connection between the merchant and buyer so that the buyer can securely approve the transaction. While this system prevents double spending and is a form of anonymity, it still relies on a central authority to monitor transactions, such as a bank [22]. It ultimately failed due to these reasons as well as nobody wanted to adopt this system. Also note that DigiCash was primarily merchant-to-user transactions, there was no support for user-to-user transactions.

To propose another issue with previous ecash systems was minting new currency. With DigiCash, to acquire $100 of DigiCash, you needed to trade in $100 of real dollars to the bank. There are of course other failed implementations of minting a digital currency, such as e-Gold, which sought to back up the digital coin with a vault of gold that was stored away. But with these implementations, the coin is still backed by some commodity or by a government. Bitcoin solves this minting and value problem by using a proof-of-work system. Here, miners use their computation resources to compete against other miners in order to obtain bitcoins in the form of a reward for solving a difficult puzzle. This difficult puzzle is the same computational puzzle as a Hashcash [23]. The reason for the difficult puzzle was security, but the reason for its success comes in the form of the reward for solving it.

The first main incentive for a node to be honest is by using a block reward [22]. The block reward is given (in bitcoin because this is currency!) to the miner who solves the Proof-of-work puzzle. The challenge here is that the transactions that the miner proposes as to be added to the blockchain must all be valid. When other miners perform their checks on the transactions, if they find that one of those transactions is invalid, those miners will choose not to extend that block onto the longest running blockchain, and thus the miner who mined that block will not get the reward. The other incentive is via transaction fees. A transaction fee is the implicit difference between the amount of the amounts and the total outputs. These transaction fees are paid to the miner who found the next block and is proposing said transactions. The transaction fee is paid by the person who is spending the bitcoins, and overtime has become an unofficial required fee. If no fee is paid, your transaction could take a long time to get into the block chain as miners will prioritize including transactions with fees.

Countries are starting to acknowledge Bitcoin as legal tender. The United States and various European countries are starting to accept bitcoin to be exchanged into green backed currency like the U.S. Dollar and Euro. However, it still stands that most of the countries still do not allow corporations to trade directly with one another via the Bitcoin system. This is where one country is starting to change.

5.1 Direct Trading

Bitcoin in El Salvador is striking controversy with its users and the World Bank. On June 8, 2021, President Nayib Bukele passed a law to make Bitcoin a legal tender. The goal of passing this law was to save on remittances from traditional money transfers. El Salvador's past with transactional fees, the hope that Bitcoin will stop the problem is high [24]. It is important to see how Bitcoin will affect the country by looking into the background of remittances, how people viewed bitcoin in the past, and how Bitcoin's remittance compares to traditional currency.

El Salvador relies on remittance payments for 20.93% of its GDP. Since the country relies on remittance payments, President Nayib Bukele is doing whatever he can to assist his people. When taking money from the World Bank, there are fees associated with those payments. Most firms charge between 3–4% per $200 and 1–2% per $500 payment. These fees add up quickly. These demanding fees is the direct reason stated by President Nayib Bukele. With that it was clear an alternative method was needed. The option that was decided on in June 2021 was Bitcoin. It seemed like a promising new and improved form of currency. However, that was not always the case.

In 2019, the coastal town of El Zonte adopted Bitcoin. It was an incentive for shop owners to use Bitcoin as a means of payment, but a major problem occurred. There is a major obstacle that the government failed to realize. To access Bitcoin as payment over the internet, the people need access to the internet. Not all citizens of El Zonte have the internet which makes payments with Bitcoin impossible. 92% of respondents said they did not want a mandation of Bitcoin, and 93.5% said they did not want to receive salaries in Bitcoin. That is the most pressing matter, but it is not the only problem occurring. The other major issue is that fees of Bitcoin may be more than expected.

Since many stores still require greenback currency, Bitcoin must be exchanged at a Bitcoin ATM. These ATMs charge a fee. The fee is 5% for every transaction (https://athenabitcoin.com/). Noted above the average fee % on payments was only 3–4% on the average highest end. The virtual transaction of Bitcoin from one user to another is significantly less than greenback currency, the fees for using Bitcoin come out to be higher than the original fee of the greenback currency.

Even though Bitcoin is becoming a recognized currency in countries, it is still far away from becoming a major currency. Even with all the security risks involved with the currency, the fact that not everyone can obtain it is crucial. The problem of remittances, the history of rejection by civilians, and the alternative fees imposed on people continue to hold back the progress of Bitcoin. Only when all people can truly access it can Bitcoin become a viable option for civilians.

6 Processing of Cryptocurrency

All cryptocurrency transactions have a beginning. A transaction occurs when a payer sends currency to a payee. When a transaction takes place, the miner then checks the payer to ensure that they have the currency, and that the payer is not trying to double spend. The miner must perform Proof of Work where the result of the resource confirms the performance [25]. This is achieved when transactions are recorded by combining the digital signatures of each party's timestamp.

6.1 Signatures

Once done, the digital code is then broadcasted to all nodes on the network. Each node then agrees the transaction is correct and then the transaction legitimate. This puts integrity into the system, using blocks. With the bitcoin protocol, all transactions are collected into a block. A block is the item that gets broadcasted to all nodes connected on the network. The way each node verifies the block is by adding a nonce to it. This way each node uses the SHA-256 hashing function to decipher the (block + nonce) algorithm [1]. It finally ends out with (block + nonce + hash). As only one block can only be verified at a time, the amount of CPU power expended can increase proportionally. Hence why CPUs are vital to bitcoin mining.

6.2   Proof of Work

This process is called Proof of Work. The purpose of this proof is to protect all miners' transactions as well as their mining. This proof functions as the fundamental security for all miners of Bitcoin. It is crucial for the proof to work as mining is how Bitcoin generates more coins for each miner which in turn increases the value that each coin is worth. With the Proof of Work, trying to cheat the system is significantly harder. Mining works in the following steps.

1. The miner selects transactions to verify
2. The transactions are then put into a Merkle Tree
3. It then extracts the root block has from the Merkle Tree
4. Then it adds a nonce or "hashes the block header"
5. The nonce and hash are incremented until the desired result is obtained
6. Once done, this is a form of Proof of Work.

However, Proof of Work is only one form of proof. Bitcoin uses Proof of Work, but there are others like Proof of Stake and Proof of Retrievability. Proof of Stake requires that the miner must show how much currency the miner already owns in the system [26]. This shows that a miner has the money to make the transactions with. Proof of Retrievability on the other hand has it where the miner is required to show that the data that was given has been stored intact and can be recovered as well [27]. This proof is set up that a miner cannot continue with a transaction until it is shown that the translation is fully complete. It also proposes that each proof relies on the fundamental aspect of the blockchain, and why every cryptocurrency utilizes it.

6.3   Proof-of-Stake

November 25, 2014, Vitalik Buterin, a co-founder of Ethereum, writes a blogpost [28] on the benefits of proof-of-stake as a consensus mechanism and addressing some arguments against proof-of-stake. On December 1, 2020, Ethereum ships "The Beacon Chain", Ethereum's phase 0 into the transition from proof-of-work consensus to proof-of-stake consensus, known as Ethereum 2.0. On the official Ethereum site, Ethereum 2.0s vision of moving from proof-of-work to proof-of-stake is to improve Ethereum's scalability, security, and sustainability. In early 2022, Ethereum plans to merge the Beacon Chain, a set of upgrades and the proof-of-stake consensus mechanism, together with the main Ethereum chain, enabling staking for the rest of the network and phasing out proof-of-work.

While there are many consensus mechanisms, proof-of-work and proof-of-stake are the most prevalent. As the concerns of the proof-of-work consensus mechanism are brought to the forefront of the cryptocurrency sphere, Ethereum has had switching to proof-of-stake in its future since early in its development. Proof-of-stake is a consensus mechanism where instead of using computational power to add blocks to the blockchain, users must stake their own tokens in order to participate in the validation of new transactions and updating the blockchain which is overseen by the network. Benefits of proof-of-stake include better energy efficiency, lower barriers to entry with reduced hardware requirements, stronger immunity to centralization, and faster transaction confirmation speeds [29].

6.4 Hashcash

Hashcash is a mechanism for providing a difficult puzzle to solve to a client. By challenging a client to solve a difficult puzzle, it provides a way to show proof-of-work, i.e that a client performed some expensive computation where the work is done in cpu cycles. Upon solving the puzzle, the client is granted a token to which they can spend. The original implementation of Hashcash was to solve the problems of spam emails [30].

Hashcash provides an efficient way to block email spammers by offering this challenge to each email they would send. For a regular user, sending a few emails a day does not provide much delay in sending the emails. But for a spammer who may be sending thousands of emails per second, solving all these challenges becomes difficult thus providing a bottleneck.

Bitcoin has adapted this concept for its proof-of-work, where the miner needs to solve a difficult challenge in order to propose the next block for consensus. In Hashcash, a long string, which consists of a version, difficulty, timestamp, resource, and random string [30]. Once the string is generated, it is hashed with a nonce value that is "guessed" by the miner. If the hashed value begins with n number of zero bits (where n is the difficulty) then the puzzle is solved. If the value does not meet this requirement, another nonce is tried until the puzzle is solved.

Now to get into some of the properties of Hashcash that make it work. Hashcash is a cost-function. This means it is expensive to compute but easy to verify [23]. To verify, you just hash the challenge string with the nonce value, and if it meets the difficulty requirement then it's solved. Hashcash is hard to solve because the best way to solve the puzzle is with a deterministic algorithm [30]. What people mean by deterministic algorithm is that there is no efficient function or value to guess for the nonce, so the best method is by using random values. Finally, Hashcash is Trapdoor-free, meaning the server that is serving the challenges has no advantage of mining tokens [30].

6.5 Security

First off, Bitcoin accomplishes security in a variety of methods. Bitcoin employs block chain technology to be an append only, tamper proof ledger. The ledger is tamper proof because the data contained in the ledger undergoes cryptographic hashing. A cryptographic hash is a 1-way function that takes any data and produces a fixed-size output in an efficient calculation [22]. Secure hash functions provide a property called collision free. A collision free function says that nobody can feasibly find X and Y, such that X does not equal Y and the hash of X equals the hash of Y. Hash functions also need a hiding property. This hiding property states that given the hash of X, it is infeasible to find X. Since the input space is larger than the output space, the pigeonhole principle states that there exist at least two inputs that would hash to the same output value.

Finally, the hash function needs to have an avalanche effect, where any change in the input completely changes the output. If we have all three of these properties satisfied, then nobody will be able to (1) change any bit of data since the hash of it will change completely; (2) nobody will be able to produce another value Y where the hash of Y equals the hash of the data, thus the data cannot be altered this way; and (3) if we hash two values X and Y together, then we can easily verify if Y is the correct hash given X and the hash value.

To quickly ensure that each transaction is immutable, we will employ the concept of a Merkle Tree [1]. In Merkle Trees, each transaction will be added to a perfect binary tree. Then from the bottom up, two pairs of transactions are hashed together. This process continues until there is one node left called the Merkle Root which contains a cryptographically sound hash of the entire tree [31]. To ensure that not a single bit is changed in each block, we need to link them together with hash pointers. With hash pointers, each block contains a pointer to the previous block's hash, thus if a single bit is changed in the block, it will break the next block's previous hash pointer.

Now that we have hash functions to provide security in the form of immutability, we need a secure way of making payments to other entities. A couple of challenges that need to be solved with a digital currency include: how can someone that is sending the currency digitally sign-off on the transaction? Where exactly does, the currency get sent to? How can we detect/prevent other malicious entities from forging our payments to steal our currency? All these challenges can be solved with public and private keys [22].

Under a digital signature algorithm, such as the Elliptic Curve Digital Signature Algorithm, both the public and private keys are derived as a pair. The public key acts as a public address which is distributed to the public. Although Bitcoin is an anonymous currency, the public key is what acts as the identity. The private key is kept securely from everyone else and is used to digitally sign messages. Once the message is digitally signed with the private key, only given the private key will the message signature be valid. A message should be easily verifiable by anyone given the public key, the message, and the signature.

How does bitcoin achieve consensus between competing nodes in the network and how does the network ensure that nodes are incentivized to be honest? Both problems can be solved with proof-of-work [22]. As described earlier, Bitcoin uses Hashcash for its proof-of-work mechanism. Once a miner successfully solves a Hashcash puzzle, they are awarded Bitcoin and collect the transaction fees from all the transactions contained in the block. After the puzzle is solved, the block is also proposed to the rest of the network for consensus. During consensus, all other nodes check the block to ensure it is really solved and go through each transaction to verify they are valid transactions. If a miner finds a fault with a translation or the block, the miner will not extend the blockchain with the proposed block. If enough miners reject the proposed block, the block will not make it in the longest running blockchain called the consensus chain and therefore the miner who mined the block will not get the reward.

Finally, a transaction model is required to make cryptocurrency a fully-fledged currency. In traditional banking with fiat currency, an account-based model is used. With an account-based model, users have an account with a balance. To make transactions, their accounts are debited and or credited based on the transaction amount. Bitcoin does not use a transaction model. Instead, they employ a new model called the Unspent Transaction Output (UTXO) model. A Bitcoin transaction contains a list of transaction inputs and transaction outputs. The inputs contain hashes for the previous transaction, an output index from which the transaction originates, and the signature of the input. The output contains the public key of the recipient, and the value of the transaction. The UTXO then consists of the hash of the transaction, and the index of the output. Since we are given the hash of the transaction from which the UTXO originates, and we are given the output index of said transaction, we can get the output value and where the output goes to. Bitcoin maintains a pool of all the UTXO's which are available to be spent, and given an UTXO, a user can submit a new transaction.

7   Conclusions

Bitcoin has many distinct attributes that have proposed its longevity. It has lasted for years already and continues to improve and become recognized. The research of Bitcoins history, problems, competition, and advantages has suggested that Bitcoin has significant structure that continues to improve. With the simulation that was created, Bitcoin is going to stay for a while. It is impossible to know if Bitcoin will be the currency of the future, but the data suggests that it will become a common form of legal tender that is utilized by numerous countries and corporations.

References

1. Nakamoto S (2009) Bitcoin: a peer-to-peer electronic cash system, May 2009
2. Rauchs M, Hileman G (2017) Global cryptocurrency benchmarking study. Number 201704-gcbs in Cambridge Centre for Alternative Finance Reports. Cambridge Centre for Alternative Finance, Cambridge Judge Business School, University of Cambridge
3. Yermack D (2013) Is bitcoin a real currency? An economic appraisal. Working Paper 19747, National Bureau of Economic Research, Dec 2013
4. Ahmed M, Akhter A, Rashid A, Fahmideh M, Pathan A-S, Anwar A (2022) Blockchain meets secured microservice architecture: a trustworthy consensus algorithm. In: Proceedings of the 19th international conference on wireless networks and mobile systems—WINSYS. INSTICC, SciTePress, pp 53–60
5. McMillan R (2014) The inside story of Mt Gox, Bitcoin's $460 Million Disaster. https://www.wired.com/2014/03/bitcoin-exchange/
6. Fargo S (2015) Bitcoin.com. https://news.bitcoin.com/
7. Vásquez J, Voia M, Balutel D, Henry C (2021) Bitcoin adoption and beliefs in Canada. https://www.bankofcanada.ca/2021/11/staff-working-paper-2021-60/
8. Chan S, Chu J, Nadarajah S, Osterrieder J (2017) A statistical analysis of cryptocurrencies. J Risk Financ Manag 10(2)
9. Huang DY, Aliapoulios MM, Li VG, Invernizzi L, Bursztein E, McRoberts K, Levin J, Levchenko K, Snoeren AC, McCoy D (2018) Tracking ransomware end-to-end. In: Proceedings—2018 IEEE symposium on security and privacy, SP 2018. Institute of Electrical and Electronics Engineers Inc, July 2018, pp 618–631
10. Wood J (2022) Crypto Ponzi schemes: how to identify and protect yourself from these scams. https://www.coindesk.com/learn/crypto-ponzi-schemes-how-to-identify-and-protect-yourself-from-these-scams/
11. Wright J, Anise O (2018) Don't@ me: hunting twitter bots at scale. Blackhat USA
12. Phillips R, Wilder H (2020) Tracing cryptocurrency scams: clustering replicated advance-fee and phishing websites. In: 2020 IEEE international conference on blockchain and cryptocurrency (ICBC), pp 1–8 (2020)
13. Chauhan A, Malviya OP, Verma M, Mor TS (2018) Blockchain and scalability. In: 2018 IEEE international conference on software quality, reliability and security companion (QRS-C), pp 122–128
14. Musk E (2021) You can now buy a Tesla with bitcoin. https://twitter.com/elonmusk/status/1374617643446063105?
15. O'Neill C, Huang J, Tabuchi H (2021) Bitcoin uses more electricity than many countries. How is that possible? https://www.nytimes.com/interactive/2021/09/03/climate/bitcoin-carbon-footprint-electricity.html
16. Schinckus C, Nguyen CP, Chong FHL (2020) Crypto-currencies trading and energy consumption. Int J Energy Econ Polic 10(3):355–364
17. Greenberg P, Bugden D (2019) Energy consumption boomtowns in the united states: community responses to a cryptocurrency boom. Energy Res Soc Sci 50:162–167
18. Farell R (2015) An analysis of the cryptocurrency industry
19. Gandal N, Halaburda H (2014) Competition in the cryptocurrency market. Technical report
20. Zhao W, Yang S, Luo X, Zhou J (2021) On peercoin proof of stake for blockchain consensus. In: 2021 The 3rd International Conference on Blockchain Technology, pp 129–134
21. Rauchs M, Hileman G (2017) Global cryptocurrency benchmarking study. Cambridge Centre for Alternative Finance. Cambridge Judge Business School, University of Cambridge
22. Narayanan A, Bonneau J, Felten E, Miller A, Goldfeder S (2016) Bitcoin and cryptocurrency technologies: a comprehensive introduction. Princeton University Press
23. Back A (2003) Hashcash—amortizable publicly auditable cost-functions, Dec 2003
24. Hanke S, Hanlon N, Chakravarthi M (2021) Bukele's bitcoin blunder. Studies in applied economics, vol 185. The Johns Hopkins Institute for Applied Economics, Global Health, and the Study of Business Enterprise, June 2021

25. Mukhopadhyay U, Skjellum A, Hambolu O, Oakley J, Yu L, Brooks R (2016) A brief survey of cryptocurrency systems. In: 2016 14th annual conference on privacy, security and trust (PST), pp 745–752
26. Nadal S, King S (2012) Ppcoin: peer-to-peer crypto-currency with proof-of-stake
27. Miller A, Juels A, Shi E, Parno B, Katz J (2014) Permacoin: repurposing bitcoin work for data preservation. In: 2014 IEEE symposium on security and privacy, pp 475–490
28. Buterin V (2014) Proof of stake: how i learned to love weak subjectivity. https://blog.ethereum.org/2014/11/25/proof-stake-learned-love-weak-subjectivity
29. Nguyen CT, Hoang DT, Nguyen DN, Niyato D, Nguyen HT, Dutkiewicz E (2019) Proof-of-stake consensus mechanisms for future blockchain networks: fundamentals, applications and opportunities. IEEE Access 7:85727–85745
30. Back A (2002) Hashcash-amortizable publicly auditable cost functions. Available http://www.hashcash.org/papers/amortizable.pdf
31. Merkle RC (1988) A digital signature based on a conventional encryption function. In: Advances in Cryptology—CRYPTO'87: Proceedings, vol 7. Springer, pp 369–378

Ransomware: A Threat to Cyber Smart Cities

Cole Lamers, Eric Spoerl, Garrit Levey, Nazim Choudhury, and Mohiuddin Ahmed

Abstract A staggering 550,000 ransomware attacks were recorded daily during the year of 2020 netting cybercriminals an immense amount of capital estimated to be over 1.5 trillion dollars. Owing their success to aging infrastructure and lack of defensive funding, malicious groups have been employing ransomware to entirely shut down critical infrastructure including medical equipment, gas, and oil pipelines, communication systems in the fields of healthcare, the military, and key spheres of industry around the world. All these incidents are posing a threat to smart city infrastructures. This also presents a large challenge for organizations when determining how to protect themselves from these popular and devastating attacks. Fortunately, the increase in ransomware attacks has caused a resurgence in cybersecurity, specifically, ransomware mitigation strategies and software which are designed to detect and prevent ransomware attacks before they can cause damage. In this chapter we will address the techniques and tools used to create and deploy ransomware, the historical effects of ransomware in large scale, global events, and the most effective techniques organizations can adopt to mitigate and prevent ransomware attacks. Through continual understanding of the nature of ransomware, we aim to educate end-users and organizations alike about the capabilities of ransomware as well as the protection strategies available in an effort to support the evolving and relentless fight against ransomware.

C. Lamers · E. Spoerl · G. Levey · N. Choudhury (✉)
Department of Computer Science, University of Wisconsin Green Bay, Green Bay, USA
e-mail: choudhun@uwgb.edu

M. Ahmed
School of Science, Edith Cowan University, Joondalup, Australia
e-mail: mohiuddin.ahmed@ecu.edu.au

1 Introduction

Ransomware is malicious software which encrypts a victim's data, preventing access to files, systems, or networks, until a specified amount of untraceable currency is paid as a ransom. It is typically installed via clicks on malicious links, opening an email attachment that install malware, or through internet downloads from the compromised websites [1]. Traditionally, the hackers usage of malware to lock up computer systems followed by the demand of ransom payments started as a low-level economic nuisance. Currently, the size and scale of the malevolent groups that use ransomware to extort billions of dollars from the public and private sectors and the severity of the perilous and Targeted attacks against critical infrastructure have increased significantly. It is an extremely lucrative criminal activity netting an estimated 20 billion dollars worldwide [2]. Attributing much of its success to easy distribution and large attack surface, ransomware continues to plague businesses big and small. According to recent cybersecurity report by the Palo Alto Networks [3], the average ransom demand rose over $2 M in 2021—a 144% increase from the year 2020. The average ransom payment ranged over $500 K with a top payment of $8.5 M. According to a report published by the Trend Micro research [4], a record 400% rise in new ransomware families was observed (150 new families) which indicates that this threat has not been contained and continues to gain momentum under inadequate antiviral solutions. New variants of ransomware are designed with an additional threat, exfiltration of data which goes beyond simply locking users out of files, and systems by also extracting and posting sensitive information on sites which hackers can purchase and use maliciously.

The first documented ransomware incident occurred in 1989 and targeted the healthcare industry by infecting 20,000 attendees of the World Health Organization AIDS conference with ransomware. The AIDS ransomware was developed by Joseph Popp and was named "PC Cyborg Trojan" [5]. It was created as a proof of concept to expose unknown capabilities of cryptography in computing. Popp used social engineering as a means to infect thousands. Social engineering is the act of tricking someone into divulging information or performing an action [6]. Although a very minimal ransom amount compared to today, The PC Cyborg Trojan set the stage for future ransomware variants. Ransomware titled "WannaCry" was developed in 2017 and resulted in the infection of over 200,000 computers in 150 different countries [7]. WannaCry was distributed to victims via phishing emails containing malicious links or file attachments. Phishing emails use social engineering tactics to trick users into clicking malicious links or file attachments. This ransomware variant was especially difficult to protect against due to its ability to spread to other computers without the need for human interaction. WannaCry relied on the ability to exploit two popular vulnerabilities in Windows machines: EternalBlue and DoublePulsar [8]. Both vulnerabilities, if successfully exploited, allowed remote attackers to execute arbitrary code on the machine leading to an overall system takeover. The widespread ransomware incurred nearly a billion dollars in damage to victim companies. The ransomware attack took place over a small time period of less than 9 h. It ended quickly because British security researcher, Marcus Hutchins was able to reverse engineer the ransomware and disable it by activating the kill switch.

As the United States continues to go through repetitive years of record-breaking amounts of ransomware attacks, it was only a matter of time until an eye-opening attack happened which affected thousands. On May 6th, 2021, the Colonial Pipeline, used for transporting critical oil across the United States was attacked by ransomware. This monumental attack allowed overseas cyber criminals to penetrate a major utility with significant impacts to the entirety of the United States eastern seaboard economy and also make off with 2.3 million dollars. The colonial pipeline was a significant event for ransomware publicity in recent times due to its impact on East Coast market forces. Gas prices rose due to fear of a gas and oil shortage as a result of the cyber attack. The entity responsible for the attack was "DarkSide", an eastern European ransomware group. Their ransomware explicitly targeted non-ex-Soviet countries and also particularly non-essential facilities such as hospitals, schools, and non-profit organizations, and government agencies. This contentious attack, along with their demands of 2 million dollars in Bitcoin shows the fuel that is continually adding to the fire [9]. Many worldwide ransomware actors have turned to the use of cryptocurrency when demanding ransom payments because of crypto's explosive growth in value as well as being very difficult to trace.

As a flourishing criminal industry, ransomware continues to pose threats not only to the personal and financial security of individuals but also risk the national security and human life. The problems continues to worsen in the current days as nearly every type of institutions including businesses, schools, governments, hospitals, and are regularly targeted, disrupted, and held hostage [10]. According to the National Cybersecurity and Communications Integration Center (NCCIC) of the U.S. Department of Homeland Security [7], ransomware attacks on businesses can lead to many disastrous consequences that include loss of sensitive or proprietary information, disruption to regular operations, financial losses due to systems and files restoration, and potential damage to an organization's reputation. In addition, ransomware infections of energy grid, nuclear plant, waste treatment facilities, or any number of critical assets, especially hospitals, medical centers and healthcare facilities could have devastating consequences, including human casualties. The latter are favorite targets for ransomware cybercriminals as in 2020, 560 healthcare facilities in the U.S. alone were infected by ransomware. Such attack of ransomware on these critical infrastructures has negative consequences in addition to the cost of recovery and delay in the treatment of critically-ill patients [11].

The existence of extensive cyber vulnerabilities across the different businesses and industry create potentially lucrative targets for the malicious ransom-seeking actors. This is the dominant driving force behind the significant increase in ransomware attacks against businesses and organizations. Today, ransomware threats and the risks of disruptions for the businesses are even conspicuous and potentially catastrophic. Cyber disruption caused by the ransomware is one of the greatest threats to the economy. The 37th U.S. deputy attorney general Rod Jay Rosenstein predicted in a report that the financial costs of global annual cybercrime will hit to $6 trillion in 2021. The escalated ransomware attacks on governments, school districts and healthcare organizations during the COVID19 pandemic demonstrated its danger and thus led the policy makers and cybersecurity experts to consider this as a serious threat to national security. Similar to the traditional warfare and physical conflict ransomware threats directly disrupt the economic stability and economic capability and thus impact the national security [12].

2 Key Industries

2.1 Healthcare

Healthcare organizations have become highly targeted by cyber criminals who wish to disrupt critical services and steal valuable sensitive data. A 2020 report concluded that 94% of healthcare organizations have experienced some form of data breach [13]. A data breach occurs when sensitive information is accessed by unauthorized personnel. For a hospital, a data breach compromises the personal and financial information stored digitally for patients. Healthcare organizations store thousands of patient records known as protected health information (PHI). PHI includes patient names, date of birth, social security numbers, payment information, etc. This information is crucial for healthcare organizations to provide the best quality of care to patients. Due to the sensitive nature, this information is very valuable to legitimate healthcare organizations and malicious actors alike. When cyber criminals steal PHI, they will sell it on the Dark Web. Due to providing online anonymity, the Dark Web has become a hub for cybercriminals to purchase and sell victim's sensitive information. Social Security numbers are worth up to 5 dollars each, credit and debit cards are roughly 100 dollars each, and driver's license information sells for about 20 dollars. Complete patient files containing multiple sensitive fields can sell for 1000 dollars each. Considering healthcare facilities store thousands of PHI records, a cybercriminal can make a lucrative amount of money on the Dark Web after a successful ransomware attack. Healthcare organizations have a large attack surface. During the COVID-19 pandemic of 2020, Internet-of-things.

(IoT) devices were used to track patient temperature data for use in predictive analytics to predict where virus outbreaks would occur throughout the world. IoT devices refer to physical objects that contain sensors, processing ability, software, and operating systems that collect data and exchange it over the networks. Unlike traditional technology, IoT devices require little to no user intervention making them excellent at monitoring and collecting data constantly. IoT devices are often designed for a single purpose and are widely diverse. Other uses for IoT devices in healthcare include pacemakers, drug infusion pumps, and xray machines. The adoption of IoT devices has allowed the healthcare industry to better serve its patients through continuous monitoring, accurate data collection, and medical technology that leads to more positive patient outcomes. Unfortunately, the rapid adoption of IoT devices has caused difficulty in balancing usability, performance, and safety resulting in organizations sacrificing security in exchange for ease of connectivity. The healthcare industry uses many technologies to provide critical patient care. These technologies often use outdated or unsupported operating systems or applications which contain vulnerabilities which can expose the system and the organization to cyber-attacks. Outdated systems and software are referred to as legacy. Legacy systems include physical equipment and software that no longer receive support from the vendor in the form of updates and patches. Due to a lack of support, these legacy systems are left exposed to multiple vulnerabilities and are difficult to protect. Healthcare organizations often lack the resources and funding necessary to adequately defend themselves against cyberattacks. In 2018, healthcare providers spent approximately 5% of their IT budgets on cybersecurity [14]. In comparison, banking and the financial sectors spend over 7% and retail and wholesale spend 6%. Due to the importance of maintaining the confidentiality and integrity of PHI, policies and regulations have been created to protect it. The US Department of Health and Human Services (HHS) issued the Health Insurance Portability and Accountability Act of 1996 (HIPAA), a law that requires healthcare organizations to follow and maintain strict national standards to protect sensitive patient health information. New variants of ransomware are designed with an additional threat, exfiltration of data. Exfiltration of PHI will likely result in a HIPAA breach. A breach under the HIPAA Rules is defined as, " . . . the acquisition, access, use, or disclosure of PHI in a manner not permitted under the HIPAA Privacy Rule which compromises the security or privacy of the PHI." [15]. A HIPAA breach is very costly to healthcare organizations as it results in hefty fines, damaged reputation, and large administrative overhead so it is crucial that organizations create policies and procedures which define how to safely access and store PHI.

2.2 Government

A research study was conducted to determine how effective ransomware was in the government sector of the United States. Results from the study concluded that between the years of 2016 and 2019, there were a reported 169 different attacks that targeted some type of government entity using ransomware. Most, if not all of the attacks were conducted against a "soft target" or an entity that had little to no funding to combat this type of threat to the data that was being housed [16]. To further strengthen the defense in the 2020 fiscal year, the United States spent 17.4 billion dollars on cybersecurity-related activities. Of the 17.4 billion, 8–10 billion dollars was injected into the Department of Defense to protect critical domestic infrastructure, 2 billion went to the Department of Homeland Security's Cybersecurity and Infrastructure Security Administration (CISA) [17]. CISA's main role is the protection of critical infrastructure to include the energy and food supply chains that were targeted. The United States already has an extensive system of legislature that has been created in the past to address emerging threats to cybersecurity vectors such as frequency, impact, and the sophistication of the attack on information systems.

Although these frameworks have served well to protect the country in the past, a great deal of revision is required to address new threats as they emerge. Currently there are more than 50 statutes that address various aspects of cybersecurity, be it directly or indirectly, unfortunately these statutes follow no overarching framework. Although the most recent of these laws have seen some type of revision since their creation, there has been no major cybersecurity legislation enacted since 2002.

In response to the lack of action, recent legislative proposals have focused on a multitude of issues to include national strategy and role of government, reform of the Federal Information Security Management Act, protection of critical infrastructure, information sharing and cross-sector coordination, breaches, cybercrime, electronic commerce privacy, international efforts, research, and development, and also creating a cybersecurity workforce [18]. Although no new law has been enacted since 2002, the Cybersecurity Act of 2012 introduced purposing regulatory frameworks and organization changes with the addition of a few recommendations from the House Republic task force to include incentives for improving private sector cybersecurity. This bill would go on to see a handful of refinement cycles and later debated on the Senate until it failed to achieve two cloture votes. It was only until major ransomware attacks began to occur at increasing frequency that the legislative branch was forced to refocus attention back to this topic and begin to draft major bills in 2020. The Chairman of the Subcommittee on Cybersecurity, Infrastructure Protection, and Innovation, Representative Yvette D. Clarke purposed the State and Local Cybersecurity Improvement Act. This act would facilitate the release of 500 million dollars' worth of grants to State, local, territorial, and tribal governments to strengthen their cybersecurity. The overall scope of this act is to encourage all sectors of government to heavily increase their funding for cyber security in their budgets. In addition, this act will also require the Cybersecurity and Infrastructure Agency (CISA) to develop a strategy to improve the cybersecurity of state, local, tribal, and territorial governments, identify and prepare Federal resources that can be quickly made available to these elements, and to also set a baseline objective for state and local cybersecurity efforts [19]. Furthermore, this act will require all echelons of government to develop comprehensive cybersecurity plans to guide the use of grant dollars and establish state and local cybersecurity resiliency committee. The creation of this act marks the first major push back from the federal government against cybersecurity related crimes and it is the first of many steps in the process of defending state and local networks from cyber criminals. Between 2015 and 2019, there were 947 proposed bills on cybersecurity from all 50 states in the hopes of avoiding future attacks [20]. Of the 947 bills that were proposed, only 233 were actually passed, a 24% success mark. The remaining 714 bills were either paused indefinitely, vetoed by the state governor, or failed to pass through legislation and were nixed [21] Interestingly enough, of the 947 proposed bills, 107 originated from the state of New York but only four were enacted, further highlighting the neglect towards cybersecurity related funding.

The framework in [22] outlines a 5-step program that entities operating in the Department of Defense can use to create policy and security plans going forward. The first step is for the entity to identify possible vectors in the network that ransomware can be inject to. To do this, the program recommends performing asset management for physical devices, clients, servers, data, software platforms, and applications. Doing so will ensure that the documentation on hand reflects the current inventory status and includes information about business use and stockholders. Additionally, step one outlines the need for documentation on possible infection vectors, malware propagation mechanisms and access methods for the assets that are public exposed. These assets will then have to be assigned a priority and outline in the continuity plan of the department in case of an attack. The second step of the plan is to protect the assets on the network. This can be done by performing regular backups, validations checks and further encryption of the data on hand. Mitigation of social engineering attacks, especially when attempting to prevent ransomware attacks, remains a large priority and a focus for education initiatives. The agency can also place responsibility on the users of the network to practice proper software hygiene by maintaining user awareness and avoid potentially hazardous websites on the web. The following step in the plan is to create an environment that encourages users to report social engineering events, reconnaissance activity, and ransomware-indicative network activity. Additionally, deploying robust anti-malware applications and host-based intrusion prevent software can further supplement the defense of the network. The fourth step of the program outlines the need for a response plan to be drafted in the event of an attack. This continuity plan must accurately guide incident response to protect data integrity and support business continuity. Creating communication channels in which stakeholders can safely express updates and move forward on their piece of the response is also a huge focus of this step. Ideally, the creation of the continuity plan will establish standards that the organization can adopt moving forward. The last step of the program is the recovery plan in case the continuity plan does not accurately address the threat. During this step the backup plans that already have been put into place will be enacted and data backups restored. Lost software will also be restored, and the event logs of the attack will be reported back to the appropriate levels of law enforcement. The trend with cryptocurrencies being constantly used among most hacking groups has not been unnoticed by the US government. Congress has recently passed in the infrastructure bill of 2021 to require reporting requirements for cryptocurrency transactions [?]. This is another tool that primarily focuses on domestic utilization of cryptocurrency, however as one of the world powers now attempting to monitor its use, this could likely fall into the trend of other governments following suit as well. A ransomware group cannot properly extort victims if there is a worldwide recognition that all transactions can be traced to a specific individual or entity like how bank transactions work. This is a part of the process to move toward resolving these problems in the world today related to our modern digital world.

2.3 Industrial

Supervisory Control and Data Acquisition (SCADA) systems are the predominant networks running the majority of all US based utilities and industrial systems. 80% of utilities in the US utilize SCADA systems [23]. SCADA systems predominantly run on Windows desktop operating systems. This has been proven to be a major issue when hardware, operating systems, and software are not appropriately upgraded, leading to intrusion vectors that can result in significant events such as the growing threat of ransomware attacks. SCADA systems are vital to the everyday infrastructure and operations of most major companies. Typically they are involved in factory or major companies with large scale supply chains required to operate with automated measurement processes (such as scales, temperature, pressure, voltages, input/output). These allow for verification of what is occurring in real time without requiring a person physically present to do so. Over time this grew into working over a network giving these workers the ability to check this information remotely as opposed to the physical location of the systems. This is possible by using data transfer protocols such as TCP/IP. This remote data transfer is the primary vector which ransomware vendors target within SCADA systems. Some major vulnerabilities that exist within SCADA are non-isolated networks, older hardware/software or operating systems, legacy tech and patch handling, and little to no network encryption. The affects of ransomware on SCADA systems as a whole can be temporarily catastrophic, such as when the city of Atlanta had power grids and a few servers locked down. The ransom payment demanded was 52,000 dollars; they rejected the payment and thus incurred 2.6 million dollars in recovery costs. For SCADA systems, the recommendations to prevent these issues involves isolating networks, scaling back privileges for users allowing minimal acess, latest patches, and altering credentials for imperative systems [24]. SCADA systems are typically written using low level programming languages such as C or C++. These languages require direct access to the memory allocated within the computer. The usage of these low level programming languages correlates with the trend of using buffer overflow attacks to be the predominant attacks against SCADA systems. A buffer overflow occurs when memory is written which exceeds the bounds and spills over into other memory buffers. This error may force the program into a state which allows the attacker to inject malicious code and compromise the system. Historically, most SCADA systems utilized closed off software from outside interaction meaning the only way to access them was to physically be present at the system. This has changed dramatically as IoT usage has grown. Aspects related to pipeline operations have now been able to provide status updates to an employee remotely. As networks needed to provide this info, they establish more points of contact that a ransomware vendor can investigate. In 2019, the President of the Interstate Natural Gas Association of America made a statement that they are now worried about the "threat from sophisticated, well-resourced nation state actors." [25]. This was foreseen in the Colonial pipeline with the speculation by the FBI of the DarkSide group targeting them.

2.4   Financial

The Small/Mid-sized Business (SMB) and financial sectors have been hit with the brunt of ransomware attacks and often it goes unnoticed because they're affecting small businesses. Small businesses tend to operate on an "as needed" basis and they do not have the resources to maintain complex systems related to computer code. On top of this, they also aren't as likely to be on top of best practices regarding information security. This poses issues that many SMB's have faced in the past decade of ransomware attacks [26]. The attacks on financial sectors not only delete information, but also publicly disclose data causing federal crimes to occur because of the release of private information. This tends to align more with financial institutions that protect account numbers, social security numbers, and other private information. These issues establish a major impact in the ways these ransomware vendors maximize their methodologies of extortion among the entities involved. Not only do they need to ensure that the data is safe within these systems, but they also need to remain vigilant for fear of violating federal laws relating to privacy [27].

3   Prevention and Mitigation

To promote security best practices, healthcare facilities can reduce the risk of ransomware attacks by allocating more of their IT budget towards cybersecurity. An increased budget would allow the IT department to purchase and utilize powerful security tools such as vulnerability scanners and intrusion prevention systems which can detect and respond to ransomware attacks. An increased IT support staff allows an organization more ability to monitor networks and systems and provide training to end users. The overall recommendation now in regard to addressing the ransomware issue is solely a defensive approach, promoting a mix of a proactive measures to minimize the event of an attack, and reactive measures to be prepared in case it does occur. Currently there is no panacea of protecting against ransomware. It is not just one set of malware, but a tree with varying branches and differentiation between them that even certain approaches to some can be completely ineffective against others.

3.1   Technical Controls

Technical safeguards include regular backups, firewall technology, blocking port access, adopting a least privilege principle, using virus and malware protection programs, encrypting stored and transferred data, and monitoring the network for intrusion attempts [28]. These proactive measures are not simple to implement and require both time and funding. However, the costs associated with falling victim to a ransomware or other cyber-attack far exceeds the cost of implementing the proactive measures.

3.1.1 Software

One way organizations can protect PHI even if it is exfiltrated by a ransomware attack is by using encryption and data masking. Encryption provides confidentiality by rendering data unreadable to unauthorized users. Data masking hides sensitive information such as social security numbers by replacing data with symbols or random values. The two main techniques for securing legacy systems are implementing an Intrusion Detection System (IDS) and tunnelling insecure communication protocols over a secure channel [29]. An intrusion detection system is a tool that monitors individual devices or an entire network searching for known threats or suspicious activity. There are two main techniques employed by IDS. The first is knowledge-based which determines if an attack is occurring by using a list of known attack signatures. If activity matches an attack signature, the IDS creates an alert that malicious activity is occurring. However, this technique will not detect unknown attacks. A behavior-based IDS will learn how a system normally operates and then will create an alert if behavior deviates from a known baseline. This is useful for detecting undefined attacks but may introduce a greater number of false positives depending on the sensitivity of the IDS. A network-based intrusion detection systems (NBIDS) is a device connected to the network segment of an organization which monitors traffic flow of network devices, collects data, and determines if network traffic is malicious. They are useful in detecting broad attacks on the network but are insufficient in detecting isolated attacks on individual devices. Contrary to a NBIDS, a host-based intrusion detection system (HBIDS) monitors individual devices. It uses software installed on a device to monitor log and system files, suspicious processes, and user privileges to determine if an attack is occurring. A major drawback of HBIDS is that software is required to be installed on individual devices which is a challenge for large organizations. Regardless of the type of IDS used, they both utilize two different models to detect malicious activity. The most customizable model is statisticalanomaly which creates a profile of normal behavior by analyzing log files, file/folder properties, and traffic patterns. If activity deviates from the normal profile, it indicates that an attack is occurring. This model suffers from a high false positive rate due to the difficulty of defining normal behavior on a system. Another challenge is that the normal profile on the IDS must be maintained regularly to provide accurate detection. Another model used by IDS is rule or signature-based. This model analyzes the sequence of events that take place during an attack and creates a unique signature to identify different attacks. If system activity matches one of these predefined signatures, the IDS raises an alert that an attack is occurring. This model is very easy to implement and maintain and has a low false positive rate. However, it is unable to detect unknown attacks such as zero days. Aastha Yadav and the coauthors [30] proposed the use of honeypots to overcome the shortcomings of traditional IDS solutions. A honeypot is a program, machine, or system put on a network as bait for attackers. According to the authors, a honeypot should be placed both in the demilitarized zone (DMZ) between the internet and an organization's local area network (LAN), as well as within the LAN itself. The authors' reasoning for this architecture is the honeypot in the DMZ can detect attacks originating from the external internet and the honeypot within the LAN of the organization can detect malicious activity from within the organization's network. It was pointed out by the authors that most attacks in healthcare are performed by insiders. The authors also proposed the use of a Dionaea and Kippo SSH honeypot system. Dionaea's main purpose is to download a copy of the malware requested by the attacker. Once downloaded, it sends the copy to multiple sandbox environments which analyze the submitted malware and can determine what kind of malware it is. Dionaea requires minimal user interaction, so it is a great tool for security personnel to determine which malware they are being targeted with. With this data, they can create security policies to combat these attacks. The second part of the proposed honeypot solution is Kippo. Kippo allows an attacker to remotely log into a system and execute commands in a remote shell. Kippo appears as a legitimate system to the attacker and creates a log of all interactions between the attacker and the shell. Kippo can track all activity of the attacker including what commands they execute, what usernames and passwords they attempt to gain access to, and what files they attempt to access. Security personnel can read Kippo logs to determine what data the attacker was trying to access as well as what techniques were used. They can then use this information to improve the security posture of the entire organization. The recommendation to circumvent the issues plaguing SCADA systems is for SCADA software to be converted over to the Rust programming language as it is a memory safe (meaning, no ability to perform buffer overflow) language that provides low level access to memory such as C/C++.

Another ransomware prevention tool is the use of a browser plugin that can detect and prevent ransomware attacks. For example, if a user clicked a phishing link or attachment, the plugin would analyze the website or the downloaded file, create a hash value, and then compare the calculated hash value with a database of ransomware hashes. If the hashes match, the plugin would report to the user that the file is malicious and can prevent the execution of the file which prevent the system from getting infected. Han et al. [31] pointed out the shortcomings of their design in that the plugin would not be able to detect unknown or new ransomware variants due to the reliance of a known database of hash signatures. Along with being able to detect malicious downloads, the plugin will also report to the user when a site their visiting online is not using Secure Sockets Layer (SSL) encryption meaning that all data sent to and received from the website can be viewed by unauthorized parties and sensitive information such as credit card numbers, banking information, and social security numbers could be stolen. The hope for the authors is that using the plugin will not only prevent infection from ransomware but also increase user awareness of cybersecurity and form healthy security habits from non-technical users. The first line of defense is using software within firewalls, antivirus, VPN, and operating systems. Firewalls restrict access to the internal network, antivirus can detect infections within systems, VPN should be used for encrypted communication of remote users, and operating systems should be kept up-todate with the latest patches to mitigate the chance of vulnerabilities. The need for artificial intelligence that can quickly evaluate human behavior will break the cycle of defensive strategies being one step behind the criminals. With the aid of artificial intelligence, the human factor of information security will drastically be reduced and will allow for more resources to be used for network upgrades [32]. MWR[1] has created a state-of-the-art ransomware agent called "RansomFlare" that uses dynamic (behavioral) analyses and machine learning to quickly understand ransomware [33]. Dynamic analysis-based detection technique is a system of live monitoring to determine if any processes on a network are behaving in malicious intent and initiates a termination function after the process has been flagged. After RansomFlare has gone through a complete iteration of termination, the software then initiates machine learning to understand the origin, behavior, and mapping of the process. RansomFlare has seen great results and offers and potential solution to the ransomware question.

3.1.2   Patching

Patching remains one of the main sources to prevent ransomware since it fixes existing vulnerabilities within a system or product [34]. The spread of malware can be countered through patching. The underlying vulnerabilities exploited by malware can be fixed by installing security patches that immunize the susceptible and potentially remove the malware from infected machines [35]. The desire for constant connectivity of multiple devices as well as the ability to easily manage and control them remotely has resulted in the creation and adoption of Internet-of-Things (IoT) devices. Examples of IoT devices include smart refrigerators, thermostats, security cameras, etc. IoT devices are often designed for a single purpose and are widely diverse. Due to resource constraints and difficulties in managing thousands of unique devices, traditional security mechanisms are inadequate and often contain vulnerabilities. Hewlett Packard (HP) in 2014 reported that 70% of IoT devices are exposed to attacks [36]. These vulnerabilities typically exist in the form of backdoors, hardcoded passwords, or unsecure firmware. In 2018, one survey noted that 28% of respondents stated that their organization does not scan for cyber vulnerabilities [37].

3.1.3   Backups

For backups to be effective, organizations should follow the 3-2-1 rule. Maintain three copies of data, use two different media formats to store the data, and store one of the copies off-site [38]. If an organization falls victim to a ransomware attack, they may be able to avoid paying the ransom if they have up-to-date and offline backups in place. According to an interview with a cloud backup service company [39], one of their healthcare customers lost access to 14 years' worth of files in an attack, but, because the victim had backup services, they did not have to pay the ransom and

---

[1] https://www.mwrcybersec.com/.

were able to regain access to files to continue operations. Backing up is the process of making copies of data which can be used to restore to a point in the past [40]. A backup process mainly involves two metrics, Recovery Point Objective (RPO) and Recovery Time Objective (RTO) [41]. RPO is the point in time where the system will be restored to. RTO is the time it takes to complete a restore. Backups can be incremental, differential, or full. Full backups copy the entire data set which allows for fast recovery but is very time consuming. Incremental backups back up data that changed since the previous backup which allow for fast back up creation. However, each incremental backup depends on the previous one meaning any damage or loss on one of the sets could prevent complete data recovery. Differential backups back up data that has changed since the last full backup. Differential backups restore quickly but take longer to create backups.

The backup solutions discussed fall into three categories: traditional backup, replication, and continuous data protection (CDP). Traditional backup is the preferred solution as it involves copying files on a system and then storing them on an isolated storage medium periodically which is usually daily, weekly, and monthly [42]. If the original system is infected, the system can be restored to a healthy state using the backup copy. Unlike a traditional backup, replication involves taking a snapshot of a system keeping a copy of the most recent data and storing it in a geographical area separated from the source system. Replication is not sufficient for ransomware recovery because the ransomware encrypts data on the source and will encrypt data on the replicated system almost immediately. CDP is like replication, but it can copy a much larger amount of data. The main drawback of CDP is that it requires a very large amount of storage making it a very expensive solution. Also, like replication, it cannot be used as a backup since changes on the source system quickly get replicated onto the target which means the CDP system will contain ransomware encrypted data.

Another important consideration businesses should make when designing backup strategies is where backups are stored geographically. If a server hosting normal user file storage gets infected by ransomware and has access to critical backup infrastructure, the backup infrastructure will be encrypted and become unusable for recovery. Therefore, it is crucial that backup solutions are "air-gapped" from the data source so that it will not be reachable by infected network systems [43]. An air gap provides physical and electronic separation of computing systems making them unable to communicate with each other. Of the three proposed backup methodologies, traditional backup is the most effective in combating ransomware. If a system is infected with ransomware and the company does not wish to pay the ransom, they can easily search the backup system for a time when it was free of ransomware and restore to that point in time. Jason Thomas [44] in his article on improving backup system evaluations suggested a backup solution that uses traditional backup methodologies based on full backups with incremental backups completed between the full backup . In addition to air-gapping the backup media, user access to the backup media should be very strict and follow the practice of least privilege allowing only administrative users to read and write to the backup system. Unfortunately, encryption ransomware has evolved to not only encrypt data on the original system, but also extends the encryption to any backups connected to the victim system or network.

Some more advanced encryption ransomware will obtain kernel privileges gaining complete control of the victim's operating system and allowing the ransomware to delete and destroy both local and network backups. This makes backups insufficient in protecting businesses from the devastating effects of encryption ransomware. Huang et al. [45] proposed the use of a ransomware-tolerant Solid-State Drive (SSD) called "FlashGuard" which allows quick and effective recovery from encryption ransomware without relying on backups . Hard-Disk Drives (HDD's) physically overwrite data on disks after a logical overwrite occurs which is time-intensive. Unlike HDD's, SSD's utilize out-of-place writes which will write to a free block of storage prior to erasing the original value causing increased performance in SSD's compared to HDD's. FlashGuard leverages these out-of-place writes to maintain a copy of a page that is deleted or modified which allows it to retain copies of data encrypted by ransomware. FlashGuard was tested using 1,447 ransomware samples encompassing 13 different families of ransomware. In these tests, the authors downloaded 1,447 ransomware samples, and executed them on virtual machines with protection services such as firewall, Microsoft security protection, and user account control disabled . They also gave the ransomware samples administrative privileges and enabled them to access the Internet to facilitate communications to command-and-control servers used in encryption. The authors observed that the encryption time ranged from 20 min to an hour and that most attempted to destroy backups. In testing the effectiveness of FlashGuard, it was noted that restoring the encrypted system was successful and the execution time for restoring ranged from 4.2 to 49.6 s. For comparison, the native approach takes 707.7 s. Using firmware-level data recovery, businesses can combat the advanced techniques modern encryption ransomware uses to prevent access to crucial data. The second line of defense is backups. Organizations should create backup solutions using both software and hardware and should regularly test the functionality of backups to ensure data is safe, accessible, and redundant. In the event of a ransomware attack that does not encrypt backups, devices can be restored to a state before they were infected with ransomware allowing access to critical files and systems.

3.2   Organizational Control

3.2.1   User Awareness Training

Kessler et al. [46] stated that the majority of data breaches result from employee negligence and/or carelessness surrounding information security, something that cannot be fully mended through legislative or technological remediation. Training the workforce is an effective strategy to prevent ransomware attacks. Organizations can increase organizational awareness by training their users on the indicators of a ransomware attack. Indicators may include clicking a link or opening an attachment within a suspicious email, an extreme increase in usage of the user's central processing unit (CPU), and an inability to access certain files. With proper training, a user can detect and respond to a ransomware attack effectively reducing the potential damage from spreading to an entire network to being contained to a single device. Users should be trained on how to identify phishing emails and instructed not to click on links or attachments within suspicious emails. With sufficient training, end users go from being a weak link in the organization's security, to being an asset in detecting and reporting attacks to technical personnel. In terms of workflow and communication, security personnel should conduct simulated phishing attacks to increase user awareness and test technical personnel's detecting and recovery strategies. Implement internal policies which restrict users' ability to install applications or access personal email accounts. Risk and business impact assessments should also be conducted to identify critical data and applications. Finally, security personnel should monitor network activity to proactively prevent ransomware attacks by identifying indicators of compromise (IOC) before an infection occurs. Typically spread through phishing emails, ransomware can be sent to hundreds of thousands of users within different organizations disguising itself as legitimate email attachments or links within email. Since most if not all users within an organization have access to email, all users of an organization can fall victim to a ransomware attack if they don't have the awareness required to detect and prevent attacks in the first place.

Cyber security awareness has a great dependence on human-factor psychology, which is the study of the interaction of people with machines and technology [47]. Humans are the weakest link in the chain of cyber security which means a simple mistake can cause significant damage if they fall victim to a ransomware attack [48]. Educating end users, will likely result in a decrease in the number of cyber security incidents and a large increase in the organization's aptitude to detect and respond to security incidents [49]. The end user rescue checklist is a step-by-step guide for users on what to do in the event of a ransomware infection. The four main components of the checklist are: (i) disconnect systems, (ii) verify encryption, (iii) determine the type of ransomware, and (iv) determine the appropriate response. Putting the checklist into practice, the first thing a user should do when they believe their system is infected is to disconnect the device from the network as well as turn off Wi-Fi Bluetooth and Hotspot functionality. In doing so, ransomware on the system will be unable to communicate over the network preventing it from receiving instructions from the command and-control server. Next, the user should try to access shared folders, network storage devices, attached storage media, and cloud-based storage to determine if any of the data was encrypted successfully by the ransomware. If the encryption was successful, the strain of ransomware should be determined to mitigate the effect of the attack. Currently, the most viable element of informational security to be exploited by potential attackers, is the human element. Not only this factor is the soft spot for most networks, but it is also the most unreliable and uncontrollable in information security. The root of this problem stems from the fact that this type of weakness is mainly caused by a lack of user awareness to the importance of information security or that they don't have a clear understanding of the risk of a potential attack. To combat this issue, corporations, governments, and other large entities have begun to draft and adopt information security awareness programs to better educate their workers on the field of information security.

With the expectations of better educating their subordinates in the field, corporations are aiming to reduce the risk to their networks by showing the warning signs and best practices to operate on. The creation of these awareness programs is undoubtedly the correct step in the right direction, but time will only tell if the initiatives bear any fruits. Ilirjana Veseli [50] argued that the creation of the plans must be supplemented with continual campaigns to continue to keep the human element in check. The effectiveness of the security awareness programs must be taken seriously and adapted depending on the findings of the campaign. When dealing with thousands of employees, there is no one size fits all approach is it is the responsibility of the chief information officer to create a blended approach that covers all the bases. One of the most difficult challenges that information security specialists have to combat is the usage of phishing in the form of social engineering and the potential for these to cripple their networks with ransomware. Due to the complexity of spear phishing, social engineering phishing attacks, it can be difficult for even trained or tech savvy employees to detect the scheme at hand. A further study by Jason Thomas [51] reported that an estimate of 80% of all organizations have experienced a phishing attack since transitioning into the digital era. This statistics is monolithic in nature and highlights a major problem that businesses, government, and technology sectors deal with. Because the lion shares of all phishing attacks target the user, a high amount of emphasis has been placed to empower the user to have a strong understanding in internet best practices. The author supported the claim to educate the users by highlighting several staggering statistics that give perspective to the scope of this problem. The author also argued that the two main focuses for criminals to attacker users is the usage of spear phishing to steal personal identities and capital through ransomware attacks. Uandykova et al. [52] stressed on providing education on the origin, objective, and ways that ransomware is deployed, such that a better state of readiness can be achieved. Similar to how major corporations require information technology related "safety classes", it may be necessary for a variety of different learning objects be pushed onto the public in order to promote a more robust understanding of the industry. A cognitive study [53] concluded that there are three short term factors the affects the human ability to make right decisions. These three factors are workload, stress, and vigilance.

3.2.2   Policies

Along with technological solution and proper education, policy development and management should be included in a multi-pronged approach to combat ransomware. Many entry vectors (e.g., software installation [54] and information flow within web applications [55]) allow malwares the administrator privileges via which they not only easily get the enhanced level of access needed to install backdoors, spyware, rootkits, or malicious bot software, and hide these from users . Three factors are considered important in this context [56]: (i) untrusted data or operations, (ii) critical data or operations that are vulnerable and need to be protected; and (iii) the specification of security requirements known as policy (e.g., the confidentiality or the integrity of data). Policies indicate whether certain types of operations or read/write accesses are permitted. Training users on security policies and using security guidelines is imperative along with developing and following strict security policies and guidelines across organizations and monitoring the effectiveness of security policies [57]. The Least Privilege Principle is still the most recommended security stance in terms of prevention in a network as it helps defend against unpatched systems and currently unknown vulnerabilities [58]. The Open Web Application Security Project (OWASP) [59] developed a comprehensive defense in depth based policy checklist and guide to prevent organizations from ransomware and ensure the proper procedures to deal with an actual ransomware outbreak. This include policy definition on least access, administrative rights, controlled access, file permissions, group policies, and incident response.

4 Conclusions

As discussed in this paper, ransomware has remained a powerful threat that must be addressed. While the concept of ransomware has existed for over 30 years, it has not been deployed at the scale in which we see it today. With the onset of "WannaCry" and "Petya" variants of ransomware throughout the world, the issues involving them have become more prevalent and significant with each passing day. Its ability to affect markets and stock behavior should certainly keep it on the minds of legislators and large enterprises alike. The attack vectors are routinely similar and payment patterns remain unchanged. Affecting all spheres of the world economy including government, healthcare, industrial, and financial with varying degrees of magnitude. The methodologies associated with defense against ransomware have been established, however it is still not a proper solution to the endemic situation at hand. Even up-to-date software can still contain exploits that make network environments vulnerable to attack. Backups have proven to be the most vital in terms of protecting against ransomware, so long as the are disconnected from the network at the time of the attack, and such that they are not infected as well. As the market for ransomware grows, regulation of cryptocurrencies will likely become mandatory as calls for tracing and monitoring of illegal transactions thrives.

References

1. FBI (2016) Internet crime complaint center (ic3): ransomware victims urged to report infections to federal law enforcement, Sep 2016
2. Committee on Homeland Security US House of Representatives (2021) Cyber threats in the pipeline: using lessons from the colonial ransomware attack to defend critical infrastructure: house committee on homeland security, Jun 2021
3. Palo Alto Networks (2022) The growing ransomware threat: 4 trends and insights, Mar 2022
4. Trend Labs (2016) The next tier, Dec 2016

5. Salvi MHU, Kerkar MRV (2016) Ransomware: a cyber extortion. Asian J Converg Technol (AJCT). ISSN-2350-1146, 2
6. Hadnagy C (2010) Social engineering: the art of human hacking. Wiley (2010)
7. Trautman LJ, Ormerod PC (2018) Wannacry, ransomware, and the emerging threat to corporations. Tenn L Rev 86:503
8. Akbanov M, Vassilakis VG, Logothetis MD (2019) Wannacry ransomware: analysis of infection, persistence, recovery prevention and propagation mechanisms. J Telecommun Inf Technol
9. Goodell JW, Corbet S (2022) Commodity market exposure to energy-firm distress: evidence from the colonial pipeline ransomware attack. Financ Res Lett 103329
10. Hayes K (2021) Ransomware: a growing geopolitical threat. Netw Secur 2021(8):11–13
11. Ransomware Task Force (2021) Combating ransomware. Intel Security Group
12. Wilner A, Jeffery A, Lalor J, Matthews K, Robinson K, Rosolska A, Yorgoro C (2019) On the social science of ransomware: technology, security, and society. Comp. Strateg 38(4):347–370
13. Bhuyan SS, Kabir UY, Escareno JM, Ector K, Palakodeti S, Wyant D, Kumar S, Levy M, Kedia S, Dasgupta D et al (2020) Transforming healthcare cybersecurity from reactive to proactive: current status and future recommendations. J Med Syst 44(5):1–9
14. Swasey K (2020) Insufficient healthcare cybersecurity invites ransomware attacks and sale of phi on the dark web. Center Anticip Intell Stud Res Rep
15. Sheffield JN (2020) The first word: the hipaa response to malware events, including ransomware attacks. Benefits Q 36(3):44–7
16. Liska A (2019) Early findings: review of state and local government ransomware attacks. Rec Future 10
17. Reeder JR, Hall CT (2021) Cybersecurity's pearl harbor moment: lessons learned from the colonial pipeline ransomware attack. Government Contractor Cybersecurity, Washington, DC, USA
18. Fischer EA (2013) Federal laws relating to cybersecurity: overview and discussion of proposed revisions. Library of Congress Washington DC Congressional Research Service
19. Department of Homeland Security (2021) Responding to ransomware: exploring policy solutions to a cybersecurity crisis: house committee on homeland security, May 2021
20. Ransomware guide
21. Skertic J (2021) Cybersecurity legislation and ransomware attacks in the United States, 2015–2019. PhD thesis, Old Dominion University
22. Snoke TD, Shimeall TJ (2020) An updated framework of defenses against ransomware. Technical report, Carnegie-Mellon Univ Pittsburgh, PA
23. Slay J, Miller M (2007) Lessons learned from the maroochy water breach. In: International conference on critical infrastructure protection. Springer, pp 73–82
24. Ibarra J, Butt UJ, Do A, Jahankhani H, Jamal A (2019) Ransomware impact to SCADA systems and its scope to critical infrastructure. In: 2019 IEEE 12th international conference on global security, safety and sustainability (ICGS3). IEEE, pp 1–12
25. Santa D (2018) Cyber and physical security, best practices, and industry and government engagement. Fed Energy Regul Comm. https://www.ingaa.org/File.aspx?id=36642&v=62328155
26. Fanning K (2015) Minimizing the cost of malware. J Corp Account Finance 26(3):7–14
27. Tariq N (2018) Impact of cyberattacks on financial institutions. J Int Bank Commer 23(2):1–11
28. Nifakos S, Chandramouli K, Nikolaou CK, Papachristou P, Koch S, Panaousis E, Bonacina S (2021) Influence of human factors on cyber security within healthcare organisations: a systematic review. Sensors 21(15):5119
29. Tervoort T, De Oliveira MT, Pieters W, Van Gelder P, Olabarriaga SD, Marquering H (2020) Solutions for mitigating cybersecurity risks caused by legacy software in medical devices: a scoping review. IEEE Access 8:84352–84361 (2020)
30. Yadav A, Raisurana S, Lalitha P (2017) Information security in healthcare organizations using low-interaction honeypot intrusion detection system. Int J Secur Appl 11(9):95–107
31. Han JW, Hoe OJ, Wing JS, Brohi SN (2017) A conceptual security approach with awareness strategy and implementation policy to eliminate ransomware. In: Proceedings of the 2017 international conference on computer science and artificial intelligence, pp 222–226

32. Mamedova N, Urintsov A, Staroverova O, Ivanov E, Galahov D (2019) Social engineering in the context of ensuring information security. In: SHS web of conferences, vol 69. EDP Sciences, p 00073
33. Nieuwenhuizen D (2017) A behavioural-based approach to ransomware detection. Whitepaper, MWR Labs Whitepaper
34. Richardson R, North MM (2017) Ransomware: evolution, mitigation and prevention. Int Manag Rev 13(1):10
35. Eshghi S, Khouzani MHR, Sarkar S, Venkatesh SS (2014) Optimal patching in clustered malware epidemics. IEEE/ACM Trans Netw 24(1):283–298
36. MacDermott Á, Kendrick P, Idowu I, Ashall M, Shi Q (2019) Securing things in the healthcare internet of things. In: 2019 global IoT summit (GIoTS). IEEE, pp 1–6
37. Robert Richardson and CSI Director (2008) CSI computer crime and security survey. Comput Secur Inst 1:1–30
38. Spence N, Bhardwaj MBBSN, Paul DP III (2018) Ransomware in healthcare facilities: a harbinger of the future? Perspectives in Health Information Management, pp 1–22
39. Zetter K (2016) 4 ways to protect against the very real threat of ransomware, May 2016
40. Laudon KC, Laudon JP (2004) Management information systems: managing the digital firm. Pearson Educación
41. Ateya IL, Shibwabo BK, Mugoh L (2015) Continuous data protection architecture as a strategy for reduced data recovery time
42. Evans C (2014) Backup vs replication, snapshots, CDP and data protection strategy. ComputerWeekly, Juni
43. Rahman NHA, Glisson WB, Yang Y, Choo K-KR (2016) Forensic-by-design framework for cyber-physical cloud systems. IEEE Cloud Comput 3(1):50–59
44. Thomas J, Galligher G (2018) Improving backup system evaluations in information security risk assessments to combat ransomware. Comput Inf Sci 11(1)
45. Huang J, Xu J, Xing X, Liu P, Qureshi MK (2017) Flashguard: leveraging intrinsic flash properties to defend against encryption ransomware. In: Proceedings of the 2017 ACM SIGSAC conference on computer and communications security, pp 2231–2244
46. Kessler SR, Pindek S, Kleinman G, Andel SA, Spector PE (2020) Information security climate and the assessment of information security risk among healthcare employees. Health Inf J 26(1):461–473
47. Elradi MD, Mohamed MH, Ali ME (2021) Ransomware attack: rescue-checklist cyber security awareness program. Artif Intell Adv 3(1)
48. Young H, van Vliet T, van de Ven J, Jol S, Broekman C (2017) Understanding human factors in cyber security as a dynamic system. In: International conference on applied human factors and ergonomics. Springer, pp 244–254
49. Hull G, John H, Arief B (2019) Ransomware deployment methods and analysis: views from a predictive model and human responses. Crime Sci 8(1):1–22
50. Veseli I (2011) Measuring the effectiveness of information security awareness program. Master's thesis
51. Thomas J (2018) Individual cyber security: Empowering employees to resist spear phishing to prevent identity theft and ransomware attacks. Thomas JE (2018) Individual cyber security: empowering employees to resist spear phishing to prevent identity theft and ransomware attacks. Int J Bus Manag 12(3):1–23
52. Uandykova M, Lisin A, Stepanova D, Baitenova L, Mutaliyeva L, Yüksel S, Dincer H (2020) The social and legislative principles of counteracting ransomware crime. Entrep Sustain Issues
53. Priestman W, Anstis T, Sebire IG, Sridharan S, Sebire NJ (2019) Phishing in healthcare organisations: threats, mitigation and approaches. BMJ Health Care Inf 26(1)
54. Sun W, Sekar R, Liang Z, Venkatakrishnan VN (2008) Expanding malware defense by securing software installations. In: International conference on detection of intrusions and malware, and vulnerability assessment. Springer, pp 164–185
55. Sabbouh M, Higginson J, Semy S, Gagne D (2007) Web mashup scripting language. In: Proceedings of the 16th international conference on world wide web, pp 1305–1306

56. Chang J, Venkatasubramanian KK, West AG, Lee I (2013) Analyzing and defending against web-based malware. ACM Comput Surv (CSUR) 45(4):1–35
57. Adel Hamdan Mohammad (2020) Ransomware evolution, growth and recommendation for detection. Mod Appl Sci 14(3):68
58. Ren A, Liang C, Hyug I, Broh S, Jhanjhi NZ (2020) A three-level ransomware detection and prevention mechanism. EAI Endorsed Trans Energy Web 7(26)
59. Frenz C, Diaz C (2018) Anti ransomware guide—owasp, Mar 2018

Printed in the United States
by Baker & Taylor Publisher Services